Step-by-Step Design of Large-Scale Photovoltaic Power Plants

Step-by-Step Design of Large-Scale Photovoltaic Power Plants

Davood Naghaviha
Daneshmand Engineers Co.
Isfahan, Isfahan, Iran

Hassan Nikkhajoei
United Globe Engineering Inc
Thornhill, ON, Canada

Houshang Karimi
Polytechnique Montreal
Montreal, QC, Canada

WILEY

This edition first published 2022

Registered Office
John Wiley & Sons, Inc., 111 River Street, Hoboken, NJ 07030, USA

Editorial Office
111 River Street, Hoboken, NJ 07030, USA

For details of our global editorial offices, customer services, and more information about Wiley products visit us at www.wiley.com.

Wiley also publishes its books in a variety of electronic formats and by print-on-demand. Some content that appears in standard print versions of this book may not be available in other formats.

Library of Congress Cataloging-in-Publication Data applied for:

ISBN: 9781119736561

Cover Design: Wiley
Cover Images: © chapin31/Getty Images; © Ali Kahfi/Getty Images

Set in 9.5/12.5pt STIXTwoText by Straive, Pondicherry, India

10 9 8 7 6 5 4 3 2 1

MIX
Paper from
responsible sources
FSC
www.fsc.org
FSC® C013604

This book is dedicated to all engineers and experts who practice in the field of photovoltaic power plants and to our families: Naghaviha's parents; Mina, Kayhan, Nikan and Behrad Nikkhajoei; Karimi's family.

Contents

Preface

The sun is the greatest source of energy and the root of other energy types. This fireball, the sun, was created about 4.603 billion years ago. Every second 2.4 million tons of the sun's mass is converted into energy. Solar energy can directly or indirectly be converted to other forms of energy, such as heat and electricity. Solar energy is used for water heating, space heating, drying of agricultural products, and electricity generation. The main obstacles to the use of solar energy include its intermittency and wide distribution.

Since the beginning of the past decade, the solar energy industry has grown up rapidly. The rapid growth has happened due to the advancement in the solar panel and inverter technology and the decrease of the solar equipment costs. The solar energy industry growth has been happening since a decade ago to address the world energy needs and to replace the conventional power plants. The fossil-fueled and atomic power plants have created environmental disasters by daily emission of tremendous amounts of carbon.

To provide sufficient supply for the global energy consumption, a cumulative amount of 18 TW of photovoltaic power plants should be installed. This means the solar energy industry has a long way to reach to a point where at least 10% of the world energy consumption is generated by solar plants. As statistics shows, by the end of 2020, the installed capacity of world photovoltaic plants has reached to more than 751 GW. This indicates an increase of about 18.5% from the total of 634 GW solar plants that had been installed by the end of 2019.

Due to the increasing number of photovoltaic (PV) plant installations, there is a higher demand for feasibility studies and detailed designs of large-scale PV power plants (LS-PVPPs). It is necessary to do the feasibility study and detailed design using a systematic and organized method. This book provides step-by-step design of large-scale PV plants by a systematic and organized method. Numerous block diagrams, flow charts, and illustrations are presented to demonstrate how to do the feasibility study and detailed design of PV plants through a simple approach.

This book includes eight chapters. Chapter 1 gives an overview of different applications and categories of solar energy, as well as the projections on the development of PV power plants worldwide. The current PV development shows a promising increase in the energy market investment despite the financial uncertainties during the Covid-19 pandemic. Chapter 2 presents the design requirements and the major phases in designing a LS-PVPP. In Chapter 3, feasibility studies are discussed in which the requirements for implementing a PV plant are evaluated from technical, legal, and economic points of view. Technical issues and existing standards related to grid connection points are studied in Chapter 4, where comprehensive technical studies are carried out using advanced simulation software based on the national network codes. Chapter 5 conducts a comprehensive discussion on radiometric terms, arrangement of solar resources, and main parameters of solar energy radiation. A detailed review on engineering documents and their classification for LS-PVPP projects is presented in Chapter 6. Main components of the DC subsystem and its designing criteria are provided in Chapter 7, and finally Chapter 8 categorizes and explains different sources of power loss in PV power plants. Energy yield prediction and its influencing factors are also discussed in this chapter.

Acknowledgment

We are grateful to all our teachers and university professors, covering the years from primary school to graduate school. Our sincere appreciation is for Prof. R. Iravani for his years of great advice throughout the Ph.D. program and postdoctoral fellowship.

We appreciate a life-long of support provided by our parents and wives for their love and affection.

Our special thanks go to our colleagues and friends for their encouragement during the preparation of this book and our professional career.

Acronyms

AC	Alternating Current
AFC	Approved for Construction
AL	Angular Losses
AM	Air Mass
AOI	Angle of Incidence
a-Si	Amorphous Silicon
BEP	Break Even Point
B&Q	Bill of Quantities
CCTV	Closed Circuit Television
CdTe	Cadmium Telluride
CF	Capacity Factor
CIS	Copper, Indium, Selenium
CT	Current Transformer
CZTS	Copper Zinc Tin Sulfide
DC	Direct Current
DEM	Digital Elevation Model
DHI	Diffuse Horizontal Irradiance
DIgSIlent	Digital Simulator for Electrical Network
DNI	Direct Normal Irradiation
DSM	Digital Surface Model
EHV	Extra-High Voltage
EPC	Engineering, Procurement, and Construction
ETAP	Electrical Transient Analyzer Program
ESC	Extremum Seeking Control
ETR	Extraterrestrial Radiation
ETS	Emits Extraterrestrial Spectrum
EVA	Economic Value Added
FLM	First-Line Management

FPV	Floating PV
GaAs	Gallium Arsenide
GaInP	Gallium Indium Phosphorous
GHI	Global Horizontal Irradiation
GIGS	Copper Indium Gallium Deselenide
GPR	Ground Potential Rise
GTI	Global Tilted Irradiation
HJT	Hetero-Junction Technology
HSE	Health, Safety, and Environmental
HV	High Voltage
IBC	Interdigitated Back Contact
IEA	International Energy Agency
IEC	International Electrotechnical Commission
IEEE	Institute for Electrical and Electronic Engineers
IFA	Issue for Approval
IFC	Issue for Comment
IFI	Issue for Information
IncCond	Incremental Conductance
IRR	Internal Rate of Return
I–V	Current–Voltage
LID	Light-Induced Degradation
LS-PVPP	Large-Scale Photovoltaic Power Plant
LV	Low Voltage
LVS	Low-Voltage Switchgear
MDL	Master Document List
Mono si	Monocrystalline Silicon
MPP	Massively Parallel Processing/Processor
MPPT	Maximum Power Point Tracking
MS-PVPP	Medium-Scale PV power plant
MV	Medium Voltage
MW	MegaWatt
NEC	National Electrical Code
NREL	National Renewable Energy Laboratory
NRMSE	Normalized Root Mean Square Errors
NPR	Nominal Power Ratio
NPV	Net Present Value
NZE	Net-Zero Emissions
NZE2050	New Net-Zero Emissions By 2050
OCV	Open-Circuit Voltage
O&M	Operations and Maintenance
PCC	Point of Common Coupling

PDC	Personal Digital Cellular System
PID	Potential-Induced Degradation
Poly-Si	Polycrystalline silicon
P&O	Perturbation and Observation
PR	Performance Ratio
PSCAD	Power System Computer-Aided Design
PT	Potential Transformer
PV	Photovoltaic
P–V	Power–Voltage
PVC	Polyvinyl Chloride
PVPP	Small-Scale PV Power Plant
SC	Solar Constant
SCADA	Supervisory Control and Data Acquisition
SCC	Short-Circuit Current
SLD	Single-Line Diagram
SPD	Surge Protective Device
STC	Standard Test Condition
THD	Total Harmonic Distortion
TSI	Total Solar Irradiance
UFC	Unified Facility Criteria
UFL	Under Frequency Load
UHV	Ultrahigh Voltage
UL	Underwriters Laboratories
UV	Ultraviolet
VCI	Virtual Central Inverter
VLS-PVPP	Very Large-Scale PV Power Plant
WBS	Work Breakdown Structure
XLPE	Cross-Linked Polyethylene

Symbols

A	Land area of a power plant
WP	Total output power
η_M	Solar module efficiency
G	Solar irradiance
L_F	Land factor
NPV	Difference between present values of the input and the output
B_n	Benefit at year n
N	Project life (year)
c_n	Cost at year n
i	Discount or interest rate
IRR	Average annual rate of return of a PV plant
I	Discount or interest rate
c_0	Initial investment
BEP	Break-Even Point
F	Total fixed costs
V	Variable costs per unit of power production
S	Savings or additional returns per unit of power production
I_{sc}	Short-circuit current at the measurement point
I_{max}	Maximum amount of the fundamental current component
r_0	Sun–earth distance
r	Average annual distance
H	Irradiation from global radiation on the horizontal plane
H_D	Irradiation from diffuse radiation on the horizontal plane
k_B	Shading correction factor as described (non-shaded: kB = 1)
R_B	Direct beam irradiation factor
R_D	Diffuse radiation factor
α_1	Horizon elevation in the γ direction
α_2	Facade/roof edge elevation relative to the solar generator plane
β	Inclination angle of the surface relative to the horizontal plane

ρ	Reflection factor of the ground in front of the solar generator
δ	Declination angle
ϕ	Latitude
ω	Hour angle
θ	Angle of incidence
L_{ST}	Standard meridian for the local time zone
L_{Ioc}	Location longitude
TE	Time
n	Day number
θ_Z	Zenith angle
ε	Sky clearness index
Δ	Sky brightness index
$I_{b,\,n}$	DNI
m	Air mass
I_E	Extraterrestrial irradiance
I_b	Beam irradiance on a horizontal surface
R_b	Ratio of the beam radiation on the tilted surface
$I_{d,\,tilt}$	DHI
ρ_g	Ground reflectance
γ	Solar azimuth
α	Solar altitude
D	Shading distance
L	Length of PV array
L'	Projective length of solar rays
K	Weight of PV array
φ	Latitude of the location
$ModuleV_{mp,\,\min}$	Minimum module voltage expected at the highest site temperature
V_{mp}	Rated module voltage at max power
$T_{\max}$	Temperature coefficient at maximum expected temperature
ΔT	Temperature variance between standard test condition (STC) and maximum expected temperature
$N_{s,\,\min}$	Minimum number of PV modules in series
$InverterV_{\min}$	Inverter minimum MPPT voltage
$ModuleV_{OC,\,\max}$	Maximum module voltage corrected for the lowest ambient temperature
V_{OC}	Rated open-circuit voltage of the PV module
$T_{\min}$	Temperature coefficient at minimum expected temperature
$N_{S,\,\max}$	Maximum number of PV modules in series
$InverterV_{\max}$	Inverter maximum allowable voltage
$ModuleI_{DC,\,\max}$	Maximum string current

I_{DC}	Short-circuit current of the PV module
$T_{\max}$	Temperature coefficient at maximum expected temperature
$P_{PV,\, nom}$	Rated PV installed power
G_{Th}	A nominal irradiance level
NPR	Inverter downsize coefficient (nominal power ratio)
$P_{inv,\, nom-total}$	Total power of the inverter that is required for the entire power plant
$P_{inv,\, nom-expected}$	Expected power of the inverter
S_{AC}	AC active power
η	Inverter efficiency
S_F	Safety coefficient and is generally considered between 1.2 and 1.3
I_O	Cable current carrying capacity from datasheet
$I_{O,\, NEW}$	Current carrying capacity corresponding to the installation conditions
T_F	Correction factor used for ambient temperature
G_F	Reduction factor used for more than one circuit
R	Cable resistance per meter.
N_{module}	Number of modules connected in series to a string
P_n	Total Pm, STC (nominal power PV module at STC) of modules connected in series (w)
V_{ID}	Reverse voltage of string diode
$V_{OC,\, STC}$	Maximum operating open-circuit voltage of the string
N_{string}	Number of PV modules in string
$I_{sc,\, STC}$	Short-circuit current
k_T	Ambient temperature coefficient from datasheet
k_L	Alternating load factor that is normally considered as 0.9
k_H	Derating factor for high number of adjacent fuses
I_{FD}	Fuse rated current given in the datasheet
$I_{DC,\, \max}$	Isc at maximum ambient temperature
k_{Ir}	Maximum irradiance
$I_{SC,\, MOD}$	Short-circuit current of the PV module
$T_{\max\, normal}$	Maximum permitted average temperature under normal conditions (35 °C)
$\Delta^{T}_{\max}$	Maximum allowable temperature rise that is 70 °C
T_{amb}	Ambient temperature
$I^{Arraypv}_{SCMT}$	Total short-circuit current of the array
$R(x)$	Reflectance
$A(x)$	Absorptance at the angle x
$T(x)$	Transmittance
C	A temperature-independent constant

E_{g0}	Band gap of the material extrapolated to absolute zero temperature
q	Elementary charge
$P_{in(q)}$	Output power of the PV set q, which is the dc input power of each inverter
$A_{S,\,I(q)}$	Shaded area of the PV set q, which is caused due to the shading by the front (southern) PV block
$P_{pv}(y, d, t, \beta)$	Actual output power of each PV module on year/day and at time
$P_{\max,\,STC}$	Maximum output power at the STC conditions
$P_{\max}(\mathrm{G, T})$	Maximum output power at a given irradiance G and temperature T
$P_{dc}(t)$	Inverter dc instantaneous power
$P_{ac}(t)$	AC power
$P_{loss}(t)$	Power losses
$P_{dc,\,pu}$	Per unit value of dc power
P_i	PV module power
$P_{\max,\,i}$	Module maximum power
r_{dc}	Resistance of the dc cable
V_{dc}	dc voltage at the cable terminals
$P_{PV,\,peak}$	Installed peak power under STC
$G(t)$	Global solar irradiance on the PV plane at time t
r_{ac}	AC cable resistance
$I_{ac,\,cable}$	Line current rms value
V_{LL}	rms value of the line voltage
N_{it}	Total number of inverters
N_i	Number of inverters that are connected to an individual AC cable
P_{Fe}	Core losses
$P_{Cu,\,N}$	Copper losses under the nominal operating power *SN* of each identical transformer
$S_t(t)$	Total apparent power loading of the installation at the moment t
N_{tr}	Number of identical transformers
P_{MPP}	MPP power
V_{MPP}	MPP voltage
I_{MPP}	MPP current
V_{string}	Total voltage of string
ΔT	Temperature variance between STC and minimum expected temperature.
I_{sc}	Short-circuit current at STC
ΔV	Percentage of cable voltage drop
L	Cable length
ΔP	Percentage of DC cable power loss
t	Short-circuit duration that is 1 second for MV

U_n	Rated fuse voltage
I_{SN}	Nominal current of a fuse
$I_{testing}$	Testing current of a fuse
$L_{pv,1}$	length of each PV module
$L_{pv,2}$	width of each PV module
$\eta_{1,DC}$	DC cables power loss coefficient
PL_{DC}	power-length product
$P_{m\text{-}sh}(y, d, t, \beta)$	output power of each PV module at the maximum power points (MPP)
$P_m(y, d, t, \beta)$	power that is produced by each PV module at the MPP
N_t	total number of strings in a PV installation
$P_{DC(t)}$	available DC power at the time t
G_{STC}	solar irradiance under STC
DP	a coefficient that accounts for the power reduction due to the temperature rise in cells
$\Delta\theta(°C)$	rise of the cell temperature above 25 °C

1
Introduction

1.1 Solar Energy

The source of solar energy is the sun. Solar energy can be classified as heat which is generated by electromagnetic waves, and light which is produced by photons. Solar energy is the main source of most of the other forms of energy available on the Earth. The solar energy is directly or indirectly converted into other forms of energy, e.g. electrical energy produced through photovoltaic (PV) technologies.

The most important feature of solar energy is that it is clean and does not harm the environment. In the long run, PV power plants will make a significant contribution to the supply of primary energy in all sectors including domestic, commercial, industrial, and transportation consumers. Moreover, factors such as government support, price of fossil fuels, cost of gas emissions CO_2, and costs of PV plant equipment affect the growth of PV plant installation capacity [1].

Figure 1.1 shows the energy conversion cycle. It shows that fossil fuels and the renewable energy sources such as biomass, wind and solar energy originate from the sun. The solar energy is stored in chemical bonds through photosynthesis of plants and produces fossil fuels millions of years later.

This book provides an overview of all aspects of designing a large-scale PV power plant (LS-PVPP) for the solar energy professionals and the university researchers. The book particularly focuses on the design of all equipment of a large-scale PV plant from the basic to advanced parts.

1.2 Diverse Solar Energy Applications

Solar energy is used for two groups of applications: non-power plant and power plant. Figure 1.2 shows various applications of solar energy [2]. Based on

Step-by-Step Design of Large-Scale Photovoltaic Power Plants, First Edition. Davood Naghaviha, Hassan Nikkhajoei, and Houshang Karimi.

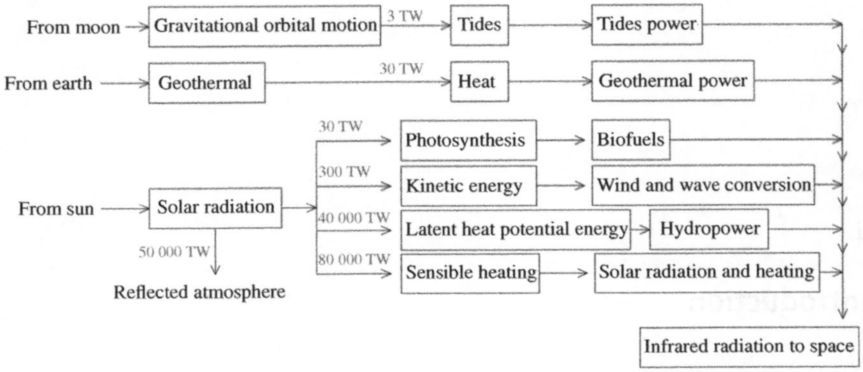
From moon
Gravitational orbital motion
3 TW
Tides
Tides power
From earth
Geothermal
30 TW
Heat
Geothermal power
30 TW
Photosynthesis
Biofuels
300 TW
Kinetic energy
Wind and wave conversion
From sun
Solar radiation
40 000 TW
Latent heat potential energy
Hydropower
50 000 TW
80 000 TW
Sensible heating
Solar radiation and heating
Reflected atmosphere
Infrared radiation to space

Figure 1.1 Energy conversion cycle. *Source:* Modified from Twidell and Weir [1].

Figure 1.2, solar power plants are divided into three categories: solar thermal; PV thermal hybrid; and PV.

1.2.1 Solar Thermal Power Plant

In a solar thermal power plant, the solar energy is converted into thermal energy which is then converted into electrical energy. Figure 1.3 shows various types of solar thermal power plants as explained below [3].

a) **Parabolic Plant**
The parabolic plant has a linear parabolic collector consisting of few rows of parabolic reflectors. The reflectors absorb the reflected rays of solar radiation and warm up the heat transfer fluid.

b) **Central Receiver Plant**
The central receiver plant consists of a set of mirrors, where each separately concentrates solar energy and transmits it to a central receiver tower.

c) **Parabolic Dish Plant**
In a parabolic dish plant, the sun's rays reflected on a parabolic surface are concentrated at a focal point. The thermal energy is converted into mechanical energy by a Stirling engine. An electric generator converts the mechanical energy into the electrical energy.

d) **Solar Chimney Plant**
In a solar chimney plant, a combination of solar air collectors and air conduction towers are used to produce induced air currents. The currents provide mechanical forces in order to rotate a pressure step turbine coupled to a generator to produce electricity.

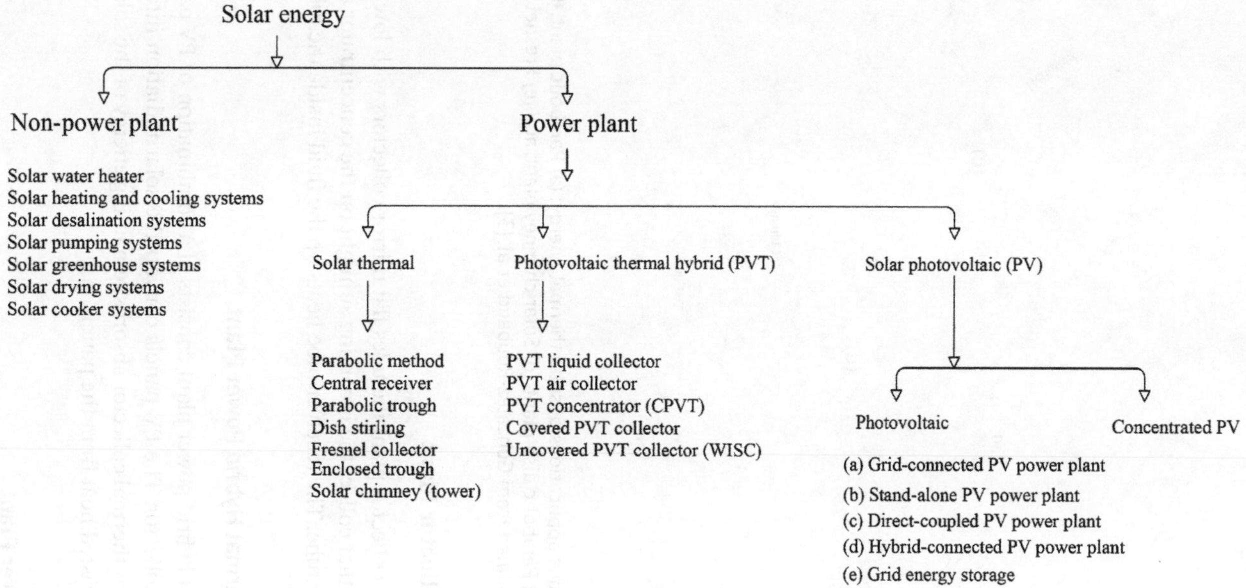

Solar energy
Non-power plant
Power plant
Solar water heater
Solar heating and cooling systems
Solar desalination systems
Solar pumping systems
Solar greenhouse systems
Solar drying systems
Solar cooker systems
Solar thermal
Photovoltaic thermal hybrid (PVT)
Solar photovoltaic (PV)
Parabolic method
Central receiver
Parabolic trough
Dish stirling
Fresnel collector
Enclosed trough
Solar chimney (tower)
PVT liquid collector
PVT air collector
PVT concentrator (CPVT)
Covered PVT collector
Uncovered PVT collector (WISC)
Photovoltaic
Concentrated PV
(a) Grid-connected PV power plant
(b) Stand-alone PV power plant
(c) Direct-coupled PV power plant
(d) Hybrid-connected PV power plant
(e) Grid energy storage

Figure 1.2 Various solar power plant categories. *Source:* Dincer and Abu-Rayash [2].

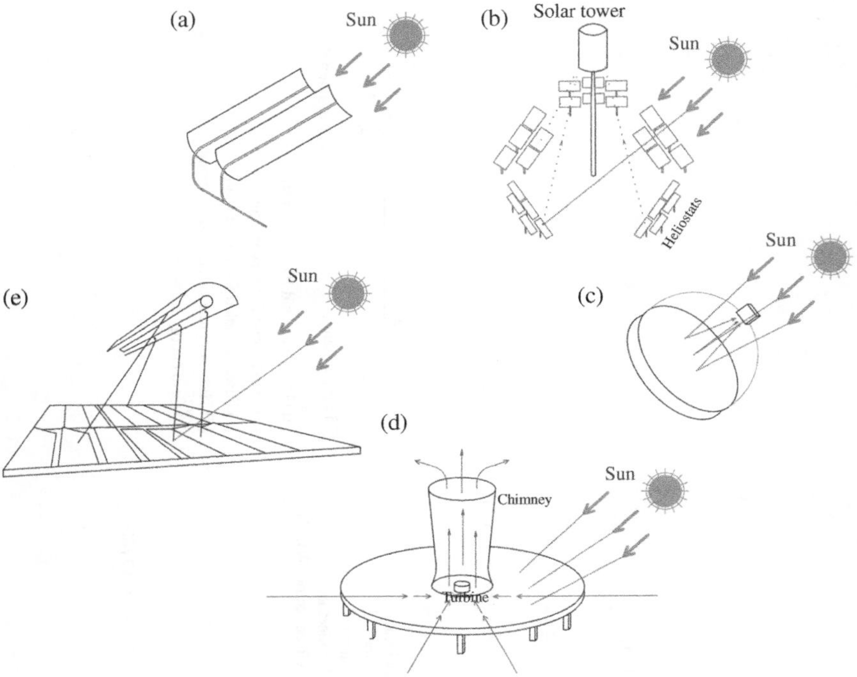
(a)
Sun
(b)
Solar tower
Sun
Heliostats
Sun
(e)
Sun
(c)
(d)
Sun
Chimney
Turbine

Figure 1.3 Various applications of solar thermal energy: (a) Parabolic plant, (b) Central receiver plant, (c) Parabolic dish plant, (d) Solar chimney plant, and (e) Fresnel collector plant. *Source:* Modified from González-Roubaud et al. [3].

e) **Fresnel Collector Plant**

The Fresnel collector plant includes flat mirror collectors with low width and long length that collect the incoming sunlight on the concentrator and send it to a receiver tube. The receiver tube heats up the fluid inside the tube.

1.2.2 PV Thermal Hybrid Power Plant

The PV thermal hybrid power plant consists of a combination of PV panels and a solar thermal collector. The PV panels convert the solar radiation into electrical energy. The solar thermal collector absorbs remaining energy of the solar rays and also removes wasted heat from the panels.

1.2.3 PV Power Plant

In a PV power plant, the sun's radiant energy is directly converted into electrical energy. There are two categories of PV power plants: conventional and concentrated.

Unlike the conventional plants, the concentrated PV plants employ curved lenses or mirrors to focus sunlight on high-efficiency PV cells. A concentrated plant has a solar tracker and a cooling system, in some cases, to further increase the plant efficiency.

Depending on the application, PV power plants are divided into five categories as briefly explained below.

a) **Grid-connected PV Power Plant**

PV power plants are usually connected to the local power network. The schematic diagram of a grid-connected PV plant is shown in Figure 1.4. For the grid-connected PV plant, the generated electricity is either consumed immediately by local loads or is sold to electricity supply companies. The local loads may include commercial and/or industrial consumers.

For the grid-connected PV plant, the grid acts as an energy storage system and, therefore, there is no need to have battery storage. In the evenings, when the PV plant is unable to produce power, the required electricity can be purchased back from the power network [4].

b) **Stand-alone PV Power Plant**

The stand-alone PV plants are used in the remote areas that have no access to the power grid. A stand-alone PV plant operates independent of the grid, with part of the produced energy stored in energy storage systems such as batteries. A schematic diagram of a stand-alone PV plant is shown in Figure 1.5. A stand-alone PV plant includes PV modules, an inverter, batteries, and a charge controller. The inverter converts the direct current generated by the PV modules to the alternating current for AC applications. The PV plant can supply both the DC and AC loads [4].

c) **Direct-coupled PV Power Plant**

In a direct-coupled PV plant, the PV array is connected directly to the load. The schematic diagram of a direct-coupled PV plant is shown in Figure 1.6. The load can operate only when there is solar radiation and, therefore, the plant

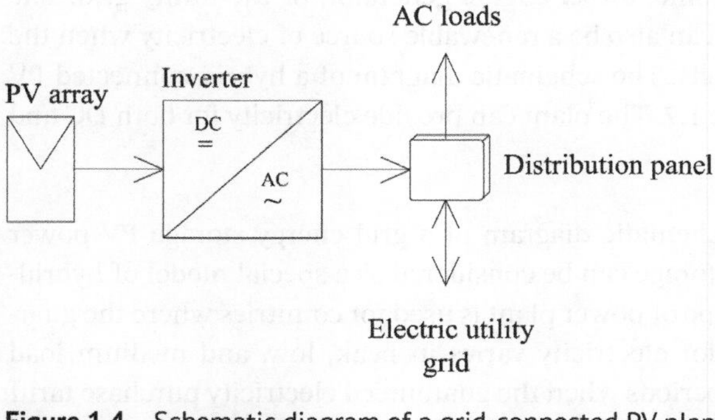
AC loads
Inverter
PV array
DC
=
AC
~
Distribution panel
Electric utility
grid

Figure 1.4 Schematic diagram of a grid-connected PV plant. *Source:* Modified from Vázquez and Vázquez [4].

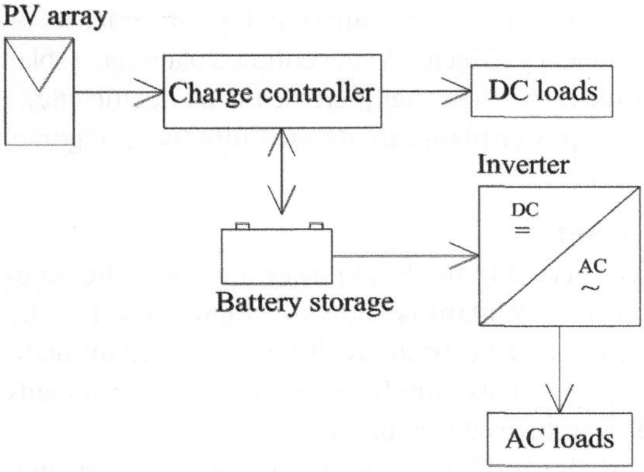
PV array
Charge controller
DC loads
Inverter
DC
=
AC
~
Battery storage
AC loads

Figure 1.5 Schematic diagram of a stand-alone PV plant. *Source:* Modified from Vázquez and Vázquez [4].

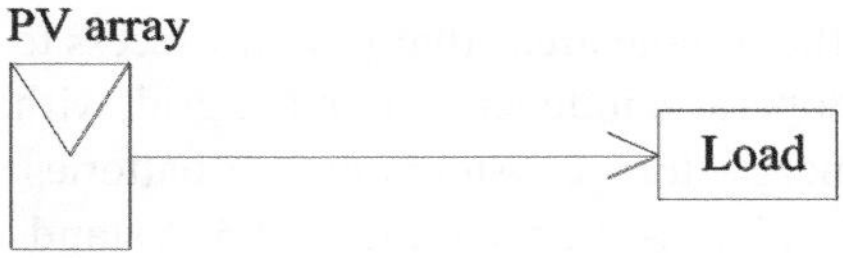
PV array
Load

Figure 1.6 Schematic diagram of a direct-coupled PV plant. *Source:* Modified from Kalogirou [5].

has limited applications. An application of this type of plant is water pumping, where the load operates as long as sunshine is available, and instead of storing the electrical energy, water is usually stored [5].

d) **Hybrid-connected PV Power Plant**

In the hybrid-connected PV plant, more than one type of generator is employed. In this type of power plant, one of the generators is a PV plant. The other generators can be wind turbine, diesel engine generator, or the utility grid. The diesel engine generator can also be a renewable source of electricity when the engine is fed with biofuels. The schematic diagram of a hybrid-connected PV plant is shown in Figure 1.7. The plant can provide electricity for both DC and AC loads [5].

e) **Grid Energy Storage**

Figure 1.8 shows the schematic diagram of a grid energy storage PV power plant. The grid energy storage can be considered as a special model of hybrid-connected plant. This type of power plant is used for countries where the guaranteed purchase tariff for electricity varies in peak, low, and medium load conditions. In the time periods when the guaranteed electricity purchase tariff

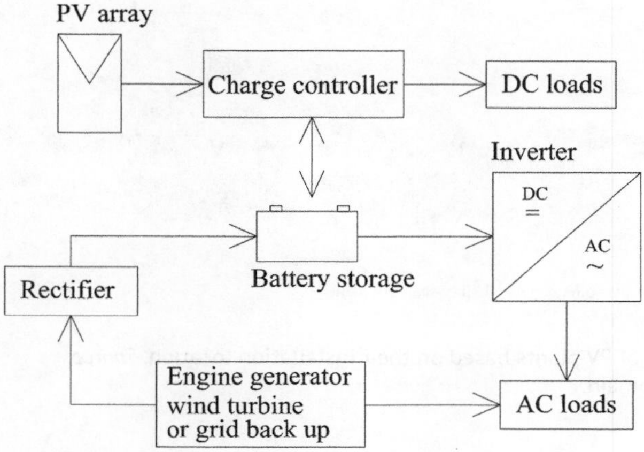
PV array
Charge controller
DC loads
Inverter
DC
=
AC
~
Battery storage
Rectifier
Engine generator
wind turbine
or grid back up
AC loads

Figure 1.7 Schematic diagram of a hybrid-connected PV plant. *Source:* Modified from Kalogirou [5].

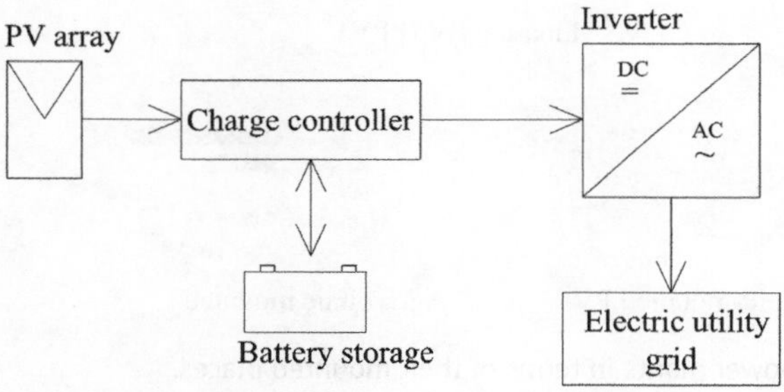
PV array
Inverter
DC
=
Charge controller
AC
~
Battery storage
Electric utility
grid

Figure 1.8 Schematic diagram of a grid energy storage PV plant.

is lower, the energy produced by the power plant is stored in batteries. When the tariff rate is higher at peak load conditions, the stored energy is injected into the grid to increase the annual revenue of the power plant.

PV power plants can be installed almost anywhere. Based on the location of installation, the PV plants are divided into three main categories: residential, industrial and commercial, and utility-scale. Figure 1.9 shows the types of PV plants based on their installation location. A residential PV system provides power to a home and/or to the grid. A commercial and industrial PV plant supply power to a corporate organization or an industrial plant. A utility-scale PV plant provides power to the grid.

Homeowners can benefit from installing a PV system in their property almost everywhere. Depending on the government policies, there are two configurations

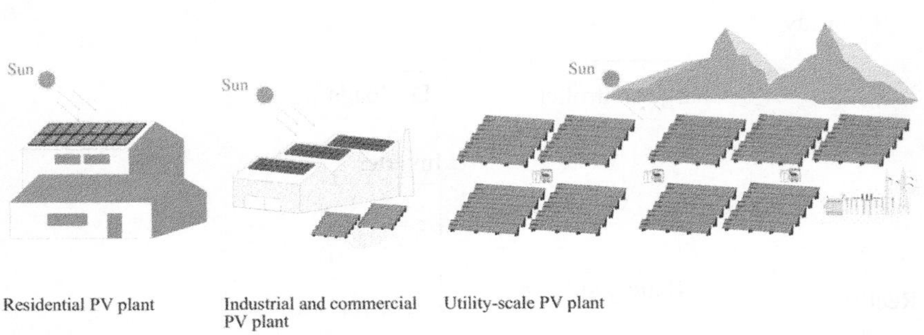
Sun
Sun
Sun
Residential PV plant
Industrial and commercial PV plant
Utility-scale PV plant

Figure 1.9 Different types of PV plants based on their installation location. *Source:* Goodrich et al. [6]. Public Domain.

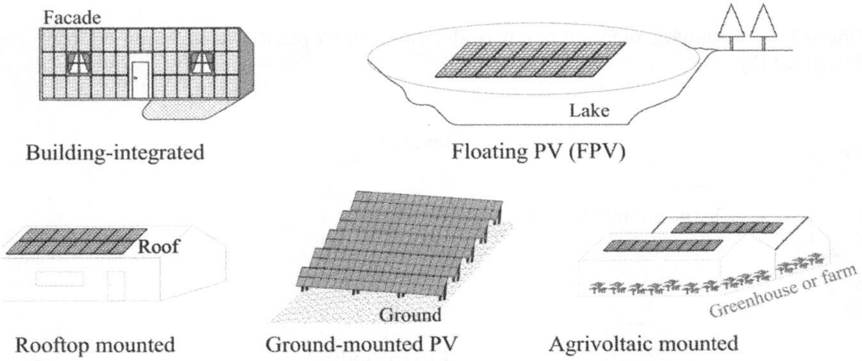
Facade
Building-integrated
Lake
Floating PV (FPV)
Roof
Rooftop mounted
Ground
Ground-mounted PV
Greenhouse or farm
Agrivoltaic mounted

Figure 1.10 Classification of power plants in terms of their mounted places.

for operating a residential PV system. In the first configuration, the residential PV system supplies the home energy, and the surplus energy is injected to the grid to offset the electricity bill through net metering. In the second configuration, the produced energy by the residential PV system is totally injected to the grid by a separate meter than the home electricity meter. The price of energy sold to the grid is paid to the home owner. Note that a utility-scale solar facility generates solar power and feeds it into the grid, i.e. supplying a utility with energy, whereas the commercial and industrial projects supply power to corporate organizations and industrial plants.

The commercial, industrial, and utility-scale PV plants are classified into five categories as shown in Figure 1.10, namely, ground-mounted, floating, building-integrated, rooftop mounted, and agrivoltaics. A ground-mounted PV plant is installed on a land, whereas a floating PV plant is installed on a water lake. In a building-integrated PV plant, solar panels are placed in the facade of a building.

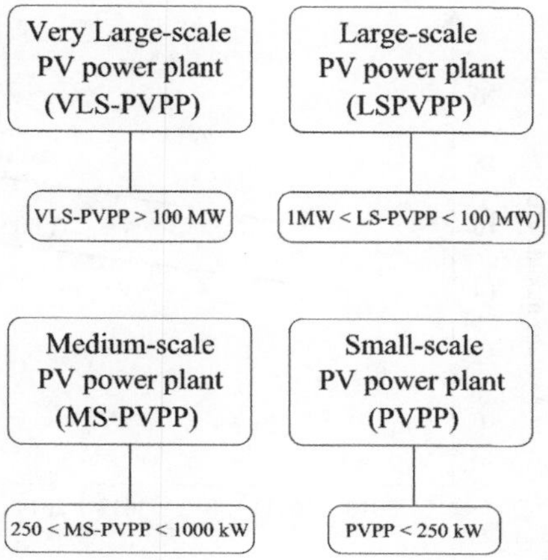
Very Large-scale
PV power plant
(VLS-PVPP)
VLS-PVPP > 100 MW
Large-scale
PV power plant
(LSPVPP)
1MW < LS-PVPP < 100 MW)
Medium-scale
PV power plant
(MS-PVPP)
250 < MS-PVPP < 1000 kW
Small-scale
PV power plant
(PVPP)
PVPP < 250 kW

Figure 1.11 Classification of power plants in terms of size. *Source:* Rakhshani et al. [8]. Licensed Under CC BY 4.0.

An agrivoltaics or agrophotovoltaics plant is installed in a greenhouse or agricultural farm and a rooftop mounted PV plant is installed on roof of a house, building, or factory [7].

Figure 1.11 shows that the PV power plants can be categorized into four groups based on their output power: small-scale, medium-size, large-scale, and very LS-PVPPs [8]. The large-scale PV plants are known as solar farms and the very large-scale PV plants are commonly known as solar parks. In addition to a distribution substation, the large-scale and very-large-scale PV plants usually have one or more transmission or sub-transmission substations.

1.3 Global PV Power Plants

In the last two decades, significant numbers of PV power plants have been installed worldwide. The cumulative installed capacity of PV plants by the end of 2020 has reached about 751 GW. There are few reasons for investing in solar plants. The most important ones are:

1) The economic incentives that some countries grant the investors by providing subsidies or purchasing PV power at high rates. Such incentives are included in the guaranteed electricity purchase tariff and are defined as a refund for the installed PV plants.
2) The improvement in the efficiency of PV modules in recent years. Figure 1.12 shows the curves of the efficiency improvement of PV modules for monocrystals and polycrystals from 2010 to 2020 [9].
3) The reduction in the costs of inverters and PV modules in recent years. The number of manufacturers of inverters and PV modules has significantly

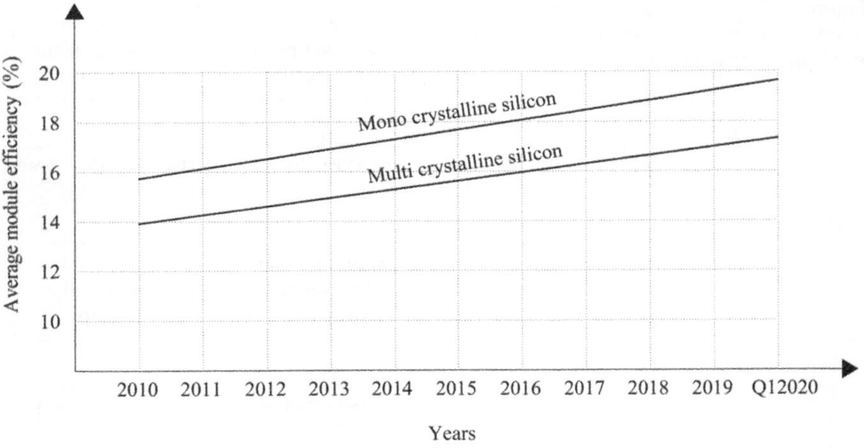

Average module efficiency (%)
20
18
16
14
12
10
Mono crystalline silicon
Multi crystalline silicon
2010 2011 2012 2013 2014 2015 2016 2017 2018 2019 Q12020
Years

Figure 1.12 Average module efficiency, 2010–2020. *Source:* Modified from Feldman et al. [9].

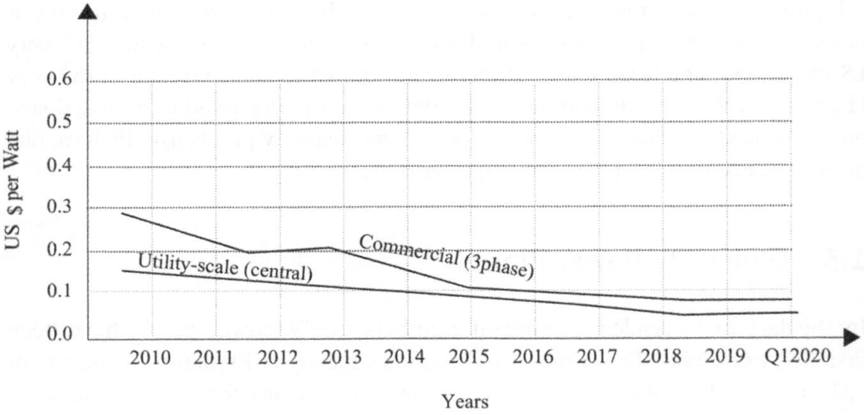

US $ per Watt
0.6
0.5
0.4
0.3
0.2
0.1
0.0
Commercial (3phase)
Utility-scale (central)
2010 2011 2012 2013 2014 2015 2016 2017 2018 2019 Q12020
Years

Figure 1.13 Utility-scale and commercial inverter prices, 2010–2020. *Source:* Modified from Feldman et al. [9].

increased over the past few years. Due to the competitive market, the price of solar inverters and PV modules has been declining to date. Reducing the costs of equipment has led to generating more economical energy from PV plants. Figures 1.13 and 1.14 show the cost trends of solar inverters and PV modules from 2010 to 2020, respectively [9].

4) Producing electric power from the sun is viable in most places and it helps to reduce the earth's greenhouse gases. Following the Paris Climate Accords,

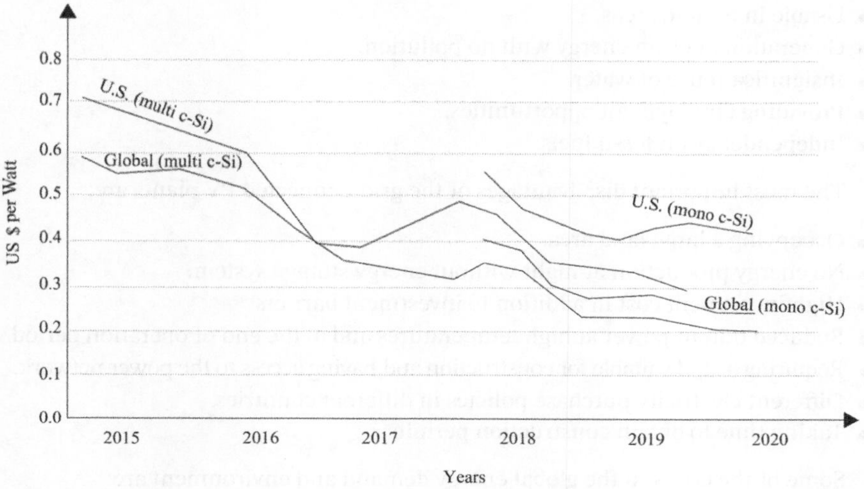
0.8
0.7
0.6
0.5
0.4
0.3
0.2
0.1
0.0
US $ per Watt
U.S. (multi c-Si)
Global (multi c-Si)
U.S. (mono c-Si)
Global (mono c-Si)
2015
2016
2017
2018
2019
2020
Years

Figure 1.14 PV modules prices, 2015–2020. *Source:* Modified from Feldman et al. [9].

adopted in December 2015, an effective way taken by most countries to reduce the greenhouse gases has been to construct more numbers of PV plants. The Paris Agreement aims to achieve a neutral climate without greenhouse gas emissions by 2050.

1.4 Perspective of PV Power Plants

To predict the growth of large-scale PV plants, the advantages and disadvantages of PV plants should be identified. A more realistic vision about the future of PV plants can be imagined by considering the fundamental crises of global energy consumption and the leading policies for its resolution. In this section, some of the advantages and disadvantages of the grid-connected PV plants are discussed.

The most important advantages of the grid-connected PV plants are:

- Worldwide availability.
- Long life of 20–30 years.
- Low depreciation due to the lack of mechanical and thermal machinery.
- Ability to generate on-site consumption and save on energy transmission and distribution costs.
- Low installation time (less than two years).
- Low maintenance costs and ease of operation.
- No production costs including the costs of fuel, marketing, and wages.

- Usable in remote areas.
- Generation of clean energy with no pollution.
- Insignificant use of water.
- Providing employment opportunities.
- Independency on fossil fuels.

The most important disadvantages of the grid-connected PV plants are:

- Occupying a large land area.
- No energy production at night without energy storage system.
- High investment cost in addition to investment barriers.
- Reduced output power at high temperatures and at the end of operation period.
- Requiring a land suitable for construction and having access to the power network.
- Different electricity purchase policies in different countries.
- Taking time to obtain construction permits.

Some of the crises in the global energy demand and environment are:

- Annual increase in energy demand.
- Limited fossil fuel resources.
- Increased environmental pollution from fossil fuels.
- Increased CO_2 emission in the atmosphere, causing global warming.
- Frequent and severe storms, drought, and sea level rises.

Given the PV plants advantages and disadvantages and global energy crises, it turns out that a large numbers of PV plants are needed to achieve long-term climate goals and to address the energy crises. Therefore, a significant growth of PV plants across the planet is expected in the near future.

The International Energy Agency (IEA) has forecast the global installed solar plants capacity by 2030 as shown in Figure 1.15. This prediction is based on considering various scenarios as described below [10].

- There are uncertainties during the Covid-19 pandemic, its economic and social impacts, and the policy responses. The stated policies assume that the Covid-19 is gradually brought under control in 2021, and the global economy returns to pre-pandemic levels in the same year.
- The delayed recovery scenario is designed with the same policy assumptions as in the stated policies scenario, but a prolonged pandemic causes lasting damage to economic prospects. The global economy returns to its precrisis size only in 2023, and the pandemic ushers in a decade with the lowest rate of energy demand growth since the 1930s.
- The sustainable development scenario assumes the energy system on track to achieve sustainable energy objectives in full, including the Paris Agreement, energy access, and air quality goals.

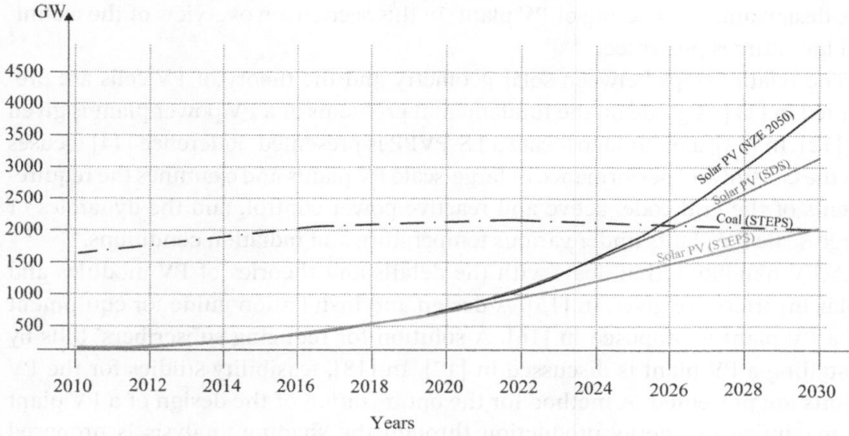
GW
4500
4000
3500
3000
2500
2000
1500
1000
500
2010
2012
2014
2016
2018
2020
2022
2024
2026
2028
2030
Years
Solar PV (NZE 2050)
Solar PV (SDS)
Coal (STEPS)
Solar PV (STEPS)

Figure 1.15 Global solar PV and coal-fired installed capacity by scenario, 2010–2030. *Source:* IEA [10].

- The new net-zero emissions by 2050 case (NZE2050) extend the sustainable development scenario analysis. A rising number of countries and companies are targeting net-zero emissions by midcentury. These are achieved in the sustainable development scenario, putting global emissions on track for net-zero by 2050. The NZE2050 includes the first detailed IEA modeling of what would be needed in the next 10 years to put global CO_2 emissions on track for net zero by 2050 [10].

Based on the IEA forecasts about the PV plant installation capacity, it can be concluded that:

- For the stated policies scenario, the PV plant installation capacity grows to about 250%.
- For the sustainable development scenario, the PV plant installation capacity increases to about 420%.
- For the NZE2050 scenario, the PV plant installation capacity grows to about 530%.

As a result, for the worst-case scenario, a PV plant installation capacity of 1250 GW is expected to be added to the existing capacity by 2030.

1.5 A Review on the Design of Large-Scale PV Power Plant

For the construction of a LS-PVPP, it is necessary to conduct preliminary studies, feasibility studies, and to obtain initial permits. Then, the most important stage is the design and engineering of PV plant. In this section, an overview of the technical literature is presented.

The relationships between solar geometry and the theory of PV cells are presented in [11]. A guide on the fundamental problems of a PV power plant is given in [12]. In [13], a method to locate a LS-PVPP is presented. Reference [14] focuses on the control and performance of large-scale PV plants and examines the requirements of the grid code, active and reactive power control, and the dynamics of large-scale PV plants under various temperature and radiation conditions.

A PV handbook that deals with the details and theories of PV modules and solar inverters are given in [15]. A design and installation guide for equipment of a PV plant is proposed in [16]. A solution for reducing subscribers' bills by installing a PV plant is discussed in [17]. In [18], feasibility studies for the PV plants are presented. A method for the optimization of the design of a PV plant to maximize its energy production through the shading analysis is proposed in [19]. Reference [20] presents a method for selecting the optimal inverter and PV modules for a PV power plant. A method for determining the cable size of a PV plant is presented in [21]. In [22], few topologies for the monitoring system of a PV plant are examined. The design of distribution transformers for a PV plant based on harmonic specifications is discussed in [23]. The protection system of a PV plant is discussed in [24, 25], and the grounding system design is presented in [26].

1.6 Outline of the Book

In Chapter 2, a review of the design requirements of a LS-PVPP is presented and various equipment of the plant is introduced. In Chapter 3, first the key points and general definitions of feasibility studies of a PV plant are introduced. Then, the criteria and requirements of a feasibility study report for a large-scale PV plant are presented.

In Chapter 4, the network connection studies of a PV power plant are discussed and the main parts of the network connection studies and its requirements are described. The single-line diagram of a sample PV plant is presented in detail. The PV plant is modeled in software, and load distribution analyzes, emergency situations, single-phase and three-phase short circuits, power quality, and stability are examined and evaluated.

In Chapter 5, first, the generalities related to solar sources, geometry, and radiation are presented. Then, a method to calculate the solar-related parameters such as total annual radiation per surface, azimuth angle, altitude angle, tilt angle and orientation, shadow distances, and row spacing is introduced.

In Chapter 6, the design methodology and documents of a LS-PVPP are presented. Moreover, the steps to design various equipment of a large-scale PV plant are discussed. The steps include preparing a feasibility study report, engineering documents, basic design, tender documents, and detailed design. Finally, a flowchart for the optimal design of a PV plant is also proposed.

The design of the DC side of a large-scale PV plant is presented in Chapter 7. The main equipment that should be determined in the DC side is introduced. The technical specifications and technologies of PV modules and solar inverters are also discussed. It is explained how to determine the PV string size, the inverter operating range, the number of inverters, the size of DC cable, and the type of fuse, surge arrester, and DC switch.

Chapter 8 introduces the power losses related to a PV plant and the parameters affecting the equipment's losses. Moreover, the performance ratio, the monthly and annual output energy productions of a PV plant are discussed.

References

1 Twidell, J. and Weir, T. (2015). *Renewable Energy Resources*. Routledge.

2 Dincer, I. and Abu-Rayash, A. (2019). *Energy Sustainability*, 75. Academic Press.

3 González-Roubaud, E., Pérez-Osorio, D., and Prieto, C. (2017). Review of commercial thermal energy storage in concentrated solar power plants: steam vs. molten salts. *Renewable and Sustainable Energy Reviews* 80: 133–148.

4 Vázquez, N. and Vázquez, J. (2018). Photovoltaic system conversion. In: *Power Electronics Handbook* (ed. M.H. Rashid), 767–781. Butterworth-Heinemann.

5 Kalogirou, S.A. (2013). *Solar Energy Engineering: Processes and Systems*. Academic Press.

6 Goodrich, A., James, T., and Woodhouse, M. (2012). Residential, commercial, and utility-scale photovoltaic (PV) system prices in the United States: current drivers and cost-reduction opportunities (No. NREL/TP-6A20-53347). National Renewable Energy Lab.(NREL), Golden, CO.

7 Jung, D., Salmon, A., and Gese, P. (2021). Agrivoltaics for farmers with shadow and electricity demand: results of a pre-feasibility study under net billing in Central Chile. *AIP Conference Proceedings* (Vol. 2361, No. 1, p. 030001). AIP Publishing LLC.

8 Rakhshani, E., Rouzbehi, K., Sánchez, A. et al. (2019). Integration of large scale PV-based generation into power systems: a survey. *Energies 12* (8): 1425.

9 Feldman, D., Ramasamy, V., Fu, R. et al. (2021). US solar photovoltaic system and energy storage cost benchmark: Q1 2020 (No. NREL/TP-6A20-77324). National Renewable Energy Lab.(NREL), Golden, CO.

10 IEA. Global solar PV and coal-fired installed capacity by scenario (2010-2030). IEA, Paris https://www.iea.org/data-and-statistics/charts/global-solar-pv-and-coal-fired-installed-capacity-by-scenario-2010-2030 (accessed 12 October 2020)

11 Fahrenbruch, A. and Bube, R. (2012). *Fundamentals of Solar Cells: Photovoltaic Solar Energy Conversion*. Elsevier.

12 White, S. (2018). *Solar Photovoltaic Basics: A Study Guide for the NABCEP Associate Exam*. Routledge.

13 Saracoglu, B.O., Ohunakin, O.S., Adelekan, D.S. et al. (2018). A framework for selecting the location of very large photovoltaic solar power plants on a global/supergrid. *Energy Reports* 4: 586–602.

14 Cabrera Tobar, A.K. (2018). Large scale photovoltaic power plants: configuration, integration and control. Doctoral thesis. Electrical Engineering Department, Universitat Politecnica de `Catalunya, Barcelona-Spain.

15 Markvart, T. and McEvoy, A. (eds.) (2003). *Practical Handbook of Photovoltaics: Fundamentals and Applications*. Elsevier.

16 Deutsche Gesellschaft für Sonnenenergie (DGS) (2013). *Planning and Installing Photovoltaic Systems: A Guide for Installers, Architects and Engineers*. Routledge.

17 Al-Najideen, M.I. and Alrwashdeh, S.S. (2017). Design of a solar photovoltaic system to cover the electricity demand for the faculty of Engineering-Mu'tah University in Jordan. *Resource-Efficient Technologies* 3 (4): 440–445.

18 Moh'd Sami, S.A., Kaylani, H., and Abdallah, A. (2013). PV solar system feasibility study. *Energy Conversion and Management* 65: 777–782.

19 Rachchh, R., Kumar, M., and Tripathi, B. (2016). Solar photovoltaic system design optimization by shading analysis to maximize energy generation from limited urban area. *Energy Conversion and Management* 115: 244–252.

20 Zidane, T.E.K., Zali, S.M., Adzman, M.R. et al. (2021). PV array and inverter optimum sizing for grid-connected photovoltaic power plants using optimization design. *Journal of Physics: Conference Series* (Vol. 1878, No. 1, p. 012015). IOP Publishing.

21 Mosheer, A.D. and Gan, C.K. (2015). Optimal solar cable selection for photovoltaic systems. *International Journal of Renewable Energy Resources* 5 (2): 28–37.

22 Ansari, S., Ayob, A., Lipu, M.S.H. et al. (2021). A review of monitoring technologies for solar PV systems using data processing modules and transmission protocols: progress, challenges and prospects. *Sustainability 13* (15): 8120.

23 Macías Ruiz, I.R., Trujillo Guajardo, L.A., Rodríguez Alfaro, L.H. et al. (2021). Design implication of a distribution transformer in solar power plants based on its harmonic profile. *Energies 14* (5): 1362.

24 Christodoulou, C.A., Ekonomou, L., Gonos, I.F., and Papanikolaou, N.P. (2016). Lightning protection of PV systems. *Energy Systems* 7 (3): 469–482.

25 Salman, S., Xin, A., Masood, A. et al. (2018). Design and implementation of surge protective device for solar panels. *2018 2nd IEEE Conference on Energy Internet and Energy System Integration (EI2)* (pp. 1–6). IEEE.

26 Nassereddine, M., Ali, K., and Nohra, C. (2020). Photovoltaic solar farm: earthing system design for cost reduction and system compliance. *International Journal of Electrical & Computer Engineering* 10 (3): 2884–2893.

2
Design Requirements

2.1 Overview

In this chapter, first, we introduce different phases of development of a large-scale photovoltaic power plant (LS-PVPP). Then, the predesign steps and the major design procedures of a large-scale solar power plant will be discussed.

Design of an LS-PVPP requires expertise in various engineering domains, technical knowledge, and experience. There are many components that should be considered in order to achieve the best performance with reasonable cost. The main demands of employers from the engineering and design services are as follows:

- Accuracy and quality of work (providing details in the executive plans presented to the contractor, appropriate to the current knowledge)
- Value engineering throughout the project
- Minimum time required to provide engineering documents
- Design guarantee after implementation

Finally, in this chapter, the main equipment needed in implementation of the LS-PVPP will be introduced.

2.2 Development Phases

Project development is the complete process of selling and preparing to install a solar PV system. The first step is marketing and sales, also known as customer acquisition or project procurement. The second step is determining the best location for a solar PV system and creating a proposal/contract. The third step is to acquire permits and to perform assessments to ensure that the system will satisfy local regulations and guidelines. Project development also includes important administrative

Step-by-Step Design of Large-Scale Photovoltaic Power Plants, First Edition. Davood Naghaviha, Hassan Nikkhajoei, and Houshang Karimi.

functions such as accounting, management, and financing. Those working in this field usually have a background in engineering (site assessor), electrical engineering (apprentice, journey worker, master), business (sales, marketing, finance, management, human resources), law with solar expertise, and project management.

Design and development of a PV project require a multidisciplinary team of experts. This is briefly described in the project development section, and then the general steps of the project are presented.

Although the main steps for developing a solar PV project are well established, there is no clear detailed "road map" that a developer can follow.

The approach taken in each project depends on site-specific parameters, developer's priorities, risk appetite, regulatory requirements, and the types of financing support mechanisms (i.e. above market rates/subsidies or tax credits) available in a given market. In all cases, certain activities need to be completed, which can broadly be organized in the following seven steps [1].

2.2.1 Concept Development and Site Identification

- Identification of potential site(s)
- Funding of project development
- Development of rough technical concept

2.2.2 Prefeasibility Study

- Assessment of different technical options
- Approximate cost/benefits
- Permitting needs
- Market assessment

2.2.3 Feasibility Study

- Technical and financial evaluation of preferred option
- Assessment of financing options
- Initiation of permitting process
- Development of rough technical concept

2.2.4 Permitting, Financing and Contracts

- Permitting
- Contracting strategy
- Supplier selection and contract negotiation
- Financing of project

2.2.5 Detailed Design and Engineering

- Preparation of detailed design for all relevant lots
- Preparation of project implementation schedule
- Finalization of permitting process

2.2.6 Construction

Construction supervision

2.2.7 Commercial Operation

- Performance testing
- Preparation of as build-design (if required)

2.3 Project Predesign

The predesign phase may include site analysis, programming, construction cost analysis, and value engineering. Site analysis includes site selection, geotechnical reports, and review of existing structures. Programming defines the project needs of the user. Programming includes cataloging the spaces, equipment needed, and functional relationships. The construction cost analysis provides a construction budget amount for the capital improvement budget and a cost plan to assist in explaining the budget and in guiding project management. Value engineering in the predesign phase scrutinizes the program, site selection, and project budget.

The following studies are also required before designing an LS-PVPP:

1) Feasibility studies
2) Grid connection studies
3) Solar resource and irradiation

2.4 Project Detailed Design

The design of an LS-PVPP involves solar technologies, solar resources, site assessment, national electrical codes, building and fire codes, local requirements, life cycle costing, etc. All these are required to create an efficient solar PV system that satisfies the expected cost and production estimates. The documentation of the design of the module substructure is intended to clearly answer the following questions:

What mounting system is used? What standards and regulations are taken into consideration? What are the local circumstances and what are the influencing factors (e.g. wind and snow loads)? What assumptions have been made during the planning step?

To answer these questions, considerable upfront efforts in planning, designing, and coordination between the customer and the mounting system provider are required [2].

It should be noted that the experts working in this sector usually have a background in:

- Science and engineering (e.g. Electrical and Power System, Structural, Civil, Environmental, Mechanical, Utility Interconnection, Computer/Software, IT specialist)
- Electrical technicians (Journey Worker, Master Electrician)
- Solar training, as well as solar industry, experience

In general, the main steps for designing an LS-PVPP are shown in Figure 2.1.

2.5 The Main Components Required for Realizing an LS-PVPP

The main electrical components of an LS-PVPP are a PV module, solar inverter, combiner box, transformer, low-voltage switchgear, medium-voltage switchgear, and high-voltage substation, as shown in Figure 2.2. The electrical components of an LS-PVPP should accomplish the following three tasks:

1) Converting solar energy into electricity
2) Connecting the LS-PVPP to the grid
3) Ensuring an adequate performance

2.5.1 PV Panels (PV Module)

Solar cells are the basis for a PV panel. The solar cells convert the solar energy into electricity. A number of solar cells are connected in series and then encapsulated in an especial frame to construct a PV panel. There are different factors that affect the overall efficiency of the PV panels, e.g. materials used in solar cells. Various solar PV technologies are described in Figure 2.3.

2.5.2 Solar Inverter

An inverter is used to convert the direct current (DC) into alternating current (AC) electricity. The output of the inverter can be single- or three-phase voltage. Inverters are rated by their total power capacity, which ranges from hundreds of watts to several megawatts [3].

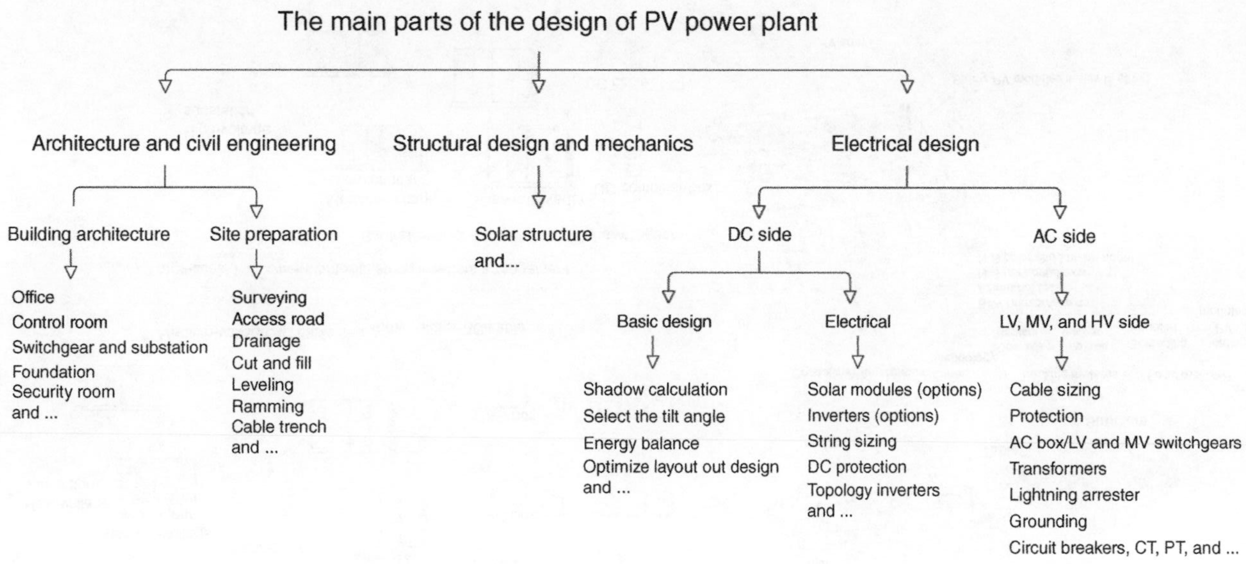
The main parts of the design of PV power plant
Architecture and civil engineering
Structural design and mechanics
Electrical design
Building architecture
Site preparation
Solar structure
and...
DC side
AC side
Office
Control room
Switchgear and substation
Foundation
Security room
and ...
Surveying
Access road
Drainage
Cut and fill
Leveling
Ramming
Cable trench
and ...
Basic design
Electrical
LV, MV, and HV side
Shadow calculation
Select the tilt angle
Energy balance
Optimize layout out design
and ...
Solar modules (options)
Inverters (options)
String sizing
DC protection
Topology inverters
and ...
Cable sizing
Protection
AC box/LV and MV switchgears
Transformers
Lightning arrester
Grounding
Circuit breakers, CT, PT, and ...

Figure 2.1 The main steps for designing an LS-PVPP.

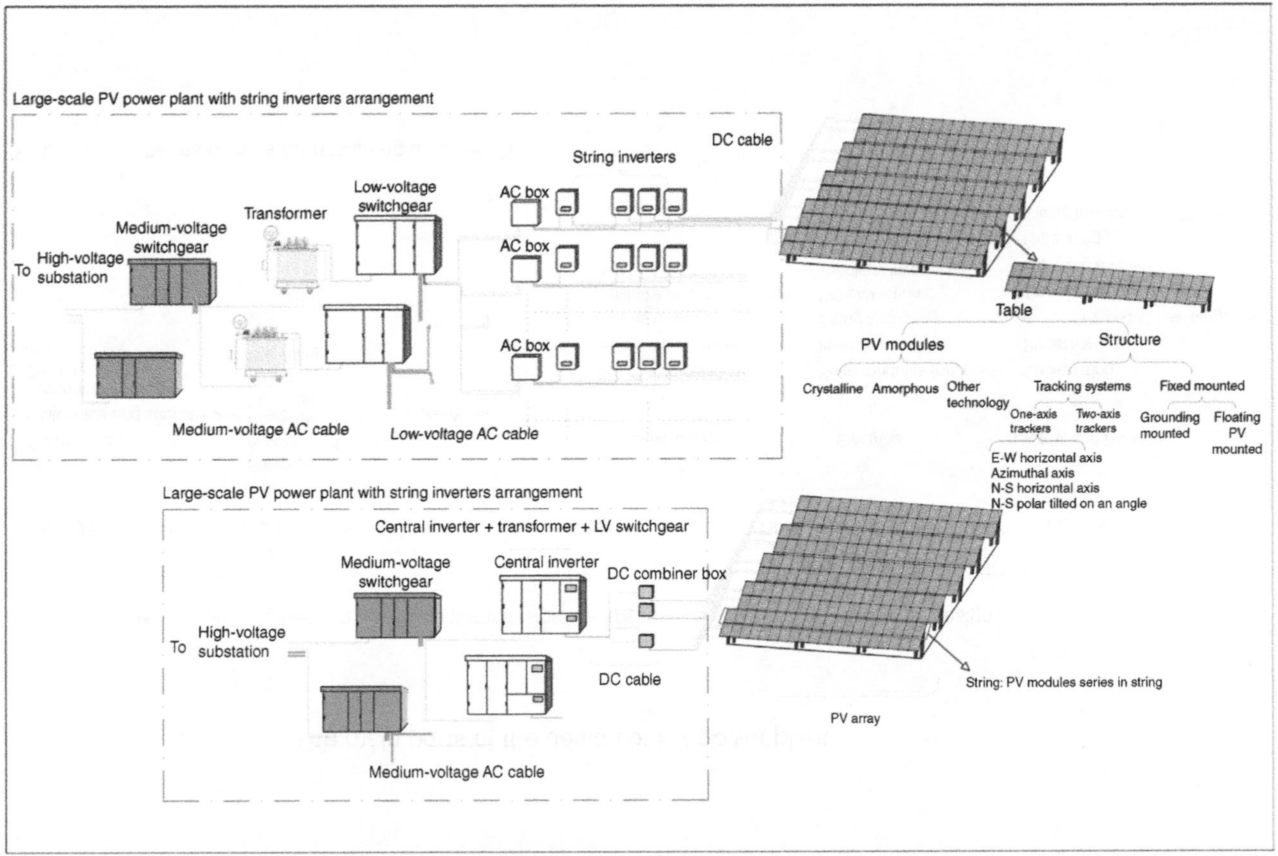
Large-scale PV power plant with string inverters arrangement
DC cable
String inverters
AC box
AC box
AC box
Low-voltage switchgear
Transformer
Medium-voltage switchgear
To High-voltage substation
Medium-voltage AC cable
Low-voltage AC cable
Table
PV modules
Structure
Crystalline
Amorphous
Other technology
Tracking systems
Fixed mounted
One-axis trackers
Two-axis trackers
Grounding mounted
Floating PV mounted
E-W horizontal axis
Azimuthal axis
N-S horizontal axis
N-S polar tilted on an angle
Large-scale PV power plant with string inverters arrangement
Central inverter + transformer + LV switchgear
Medium-voltage switchgear
Central inverter
DC combiner box
To High-voltage substation
DC cable
Medium-voltage AC cable
String: PV modules series in string
PV array

Figure 2.2 The main components needed for realization of an LS-PVPP.

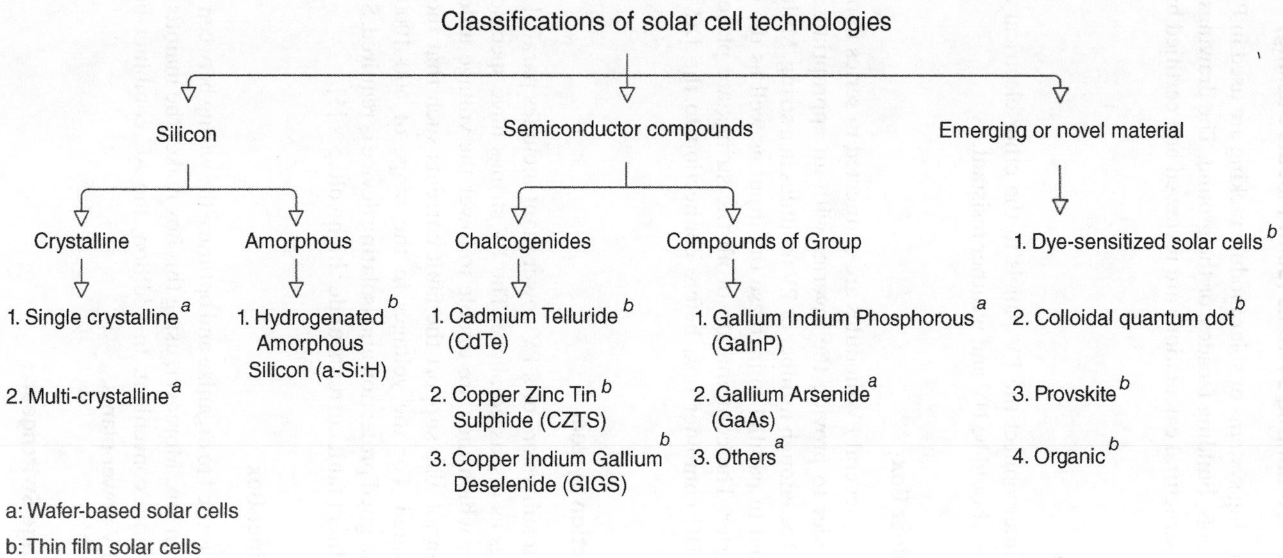

Classifications of solar cell technologies
Silicon
Semiconductor compounds
Emerging or novel material
Crystalline
Amorphous
Chalcogenides
Compounds of Group
1. Single crystalline a
2. Multi-crystalline a
1. Hydrogenated Amorphous Silicon (a-Si:H) b
1. Cadmium Telluride (CdTe) b
2. Copper Zinc Tin Sulphide (CZTS) b
3. Copper Indium Gallium Deselenide (GIGS) b
1. Gallium Indium Phosphorous (GaInP) a
2. Gallium Arsenide (GaAs) a
3. Others a
1. Dye-sensitized solar cells b
2. Colloidal quantum dot b
3. Provskite b
4. Organic b
a: Wafer-based solar cells
b: Thin film solar cells

Figure 2.3 Classification of solar cells based on the primary active material.

2.5.3 Photovoltaic Mounting Systems (Solar Module Racking)

Photovoltaic mounting systems or solar module racking are used in PV panels on surfaces such as roofs, building facades, or the ground. The drawings for mounting structure and structural calculations are reviewed and certified by a licensed engineer.

2.5.4 DC Cable

Solar DC cables interconnect the PV panels to the other electrical DC components. These cables should be UV and weather resistant.

2.5.5 DC Combiner Box

For large PV plants, several PV modules are connected in series in order to create a string. In order to provide the inverter with an appropriate DC voltage level, there should be enough number of PV modules in a string. Multiple strings are then connected in parallel to increase the input as well as the output currents of the inverter. The combination of all the strings are placed in a box, referred to as a DC combiner box, before connection to the DC side of the inverter [4].

2.5.6 DC Protection System

The direct current subsystem of a PV power plant includes parallel connection of multiple strings as discussed above. The PV strings have specific protection characteristics for which they are unable to lower the voltage unless they are obscured, and cannot also support the fault currents such that the protection devices are activated. For the voltages in the range of 600–1000 V DC and beyond, a well-designed protection and isolating device is required. Such devices must withstand direct fault currents under high voltages [5].

2.5.7 AC Combiner Box

AC combiner box is used to organize and optimize the wiring between the inverter and the distribution box. Moreover, using this box makes the maintenance of the PV power plant more convenient. In addition, the AC combiner box provides higher safety for PV power plants.

2.5.8 Low-Voltage Switchgear

Low-voltage switchgear (LVS) are devices that protect, control, and detect fault in electrical power system.

2.5.9 Transformers

Step-up transformers connect PV plants to the utility grid. Since the environmental conditions of PV plants are extremely severe, the transformers should cope with the high temperatures and harsh weather conditions. Sizing of the transformers is an important task when planning a PV power plant. Large-rated power transformers may lead to instabilities and are economically disadvantageous, whereas small-rated transformers cannot exploit the full capacity of the installed PV plant. Solar inverters transform the DC power generated from the solar modules into the AC power and feed this power into the network. A special multiple windings design of the transformer enables several PV panel strings to be connected to the grid with a minimum number of transformers [6].

2.5.10 Medium-Voltage Switchgear

Medium-voltage switchgear are devices that protect, control, and detect fault in electrical power system.

2.5.11 LV and MV AC Cables

LV and MV AC cables are used in the AC side and interconnect the solar inverters to the other electrical components.

2.5.12 AC Protection Devices

The main objective of the protection system is to ensure safe and reliable transfer of electrical power to the consumers. The protection system utilizes heavy duty and expensive equipment.

The AC protection devices should control all electrical quantities, e.g. voltage, current, and frequency. Limiting these electrical quantities guarantees the reliability of the power system protection [7].

2.6 An Overview of PV Technologies

This section briefly presents commercially available technologies of solar PV modules. Moreover, the module certification and how PV module performance can be degraded over the time are discussed.

2.6.1 Background on Solar Cell

Solar cell is a key device that converts the light energy into the electrical energy. Most PV technologies use semiconductor as their cell material. The energy conversion consists of absorbing the light (photon) energy and producing electron–hole pairs in a semiconductor. Then, the charge carrier separation occurs. A p–n junction is used for the charge carrier separation in most cases. The type of p–n junction in solar cell affects the efficiency, manufacturing cost, and energy consumption for the fabrication [8].

2.6.2 Types and Classifications

PV energy conversion is based on the principle that when a solar cell is illuminated by sunlight, it generates electricity. A number of factors including cell material (e.g. silicon, semiconductor compounds, etc.), cell size, and intensity and quality of the light source determine the amount of electricity generated. For example, the larger the cell size, the more voltage and current are produced. Therefore, the PVs are generally classified based on either the active materials used for the cells, i.e. the primary light-absorbing materials, or the overall device structures. Figure 2.3 shows the classification of solar cells based on the primary active material.

The classifications based on material complexity have been recently proposed. In terms of device structure and architecture, the PVs are categorized into wafer-based and thin-film technologies. The wafer-based PVs are produced from slices of semiconducting wafers obtained from ingots. The thin-film cells adopt an inherently different approach where the insulating substrates, such as glass or flexible plastics, are used for the deposition of layers of semiconducting materials forming the device structure.

2.7 Solar Inverter Topologies Overview

Inverters are power electronic systems that convert a DC voltage to an AC voltage/current with controllable frequency, amplitude, and phase angle. For the case of LS-PVPPs where PV panels generate DC voltage, a solar inverter interfaces the string of panels with the AC power grid. There are four types of solar inverters: central inverter; multi-string inverter; string inverter; and micro-inverter, Figure 2.4.

2.7.1 Central Inverter

In a central inverter, all module strings are merged centrally and strings are then connected in parallel to the inverter. This is particularly useful when all PV modules are subject to similar conditions in terms of inclination and orientation [9]. Central inverters are often used in large photovoltaic systems with a rated output of more than 100 kW.

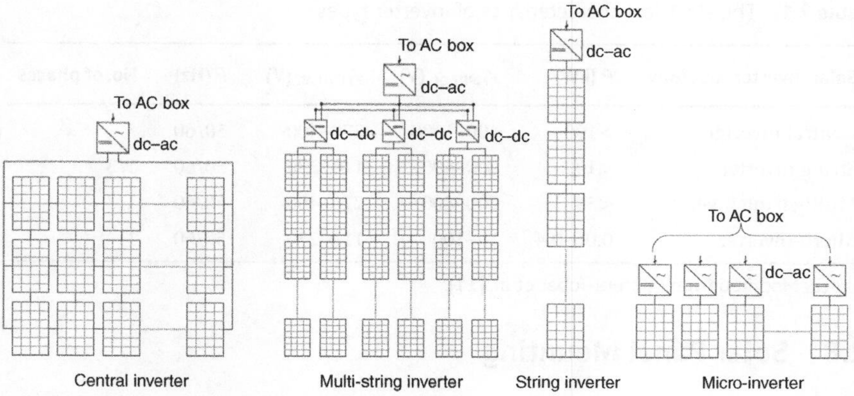
To AC box
dc–ac
To AC box
dc–ac
dc–dc
dc–dc
dc–dc
To AC box
dc–ac
To AC box
dc–ac
To AC box
dc–ac
Central inverter
Multi-string inverter
String inverter
Micro-inverter

Figure 2.4 Types of solar inverters.

2.7.2 String Inverter

A string inverter is used to convert the DC power of a string of PV modules to the AC power. It is the most common type of inverter used in residential and small/medium commercial solar systems [10]. String inverter is usually located at a short distance from the solar string in a sheltered location between the string and the switchboard.

2.7.3 Multi-string Inverter

Multi-string inverter is connected to multiple PV strings, and includes multiple DC–DC converters, each connected to a PV string. The outputs of the DC–DC converters are connected to a common DC link, where the DC side of the inverter is located [10].

2.7.4 Micro-Inverter

Micro-inverter is connected to one solar panel. Micro-inverters best suit to the places where shading is a major issue. This inverter type is also suitable for the roofs that are too small and do not provide enough space to install a string of panels [10]. Micro-inverter-based solar system is more expensive to install than other inverter configurations. Furthermore, with micro-inverters there is potentially a higher chance of failure due to the more number of inverters. However, micro-inverters have been used for years since they offer advantages with respect to the string inverters.

Table 2.1 provides the electrical characteristics of each type of inverter [11].

Table 2.1 The electrical characteristics of inverter types.

Solar inverter topology	*P* (kW)	$V_{MPPT\ DC}$ (V)	$V_{MPPT\ AC}$ (V)	*F* (Hz)	No. of phases
Central inverter	>100	400–1000	270–400	50/60	3
String inverter	<100	200–500	110–230	50/60	1/3
Multi-string inverter	<50	200–800	270–400	50/60	1/3
Micro-inverter	0.06–0.4	20–100	110–230	50/60	1

Source: Modified from Cabrera-Tobar et al. [11].

2.8 Solar Panel Mounting

There are few types of solar panel mountings. The mounting type, which is suitable for a string of solar panels, depends on few factors including the panels' location, amount of available solar radiation, and costs [12]. The mounting types include the following.

- **Roof-mounted Panels**: This type is suitable for shingle roofs with low slope.
- **Building-integrated Panels**: Panels are mounted on the walls and windows of a building.
- **Ground-mounted Panels**: Panels are mounted on metal structures on the ground. The mounting structure should comply with the standard UFC 3-301-0.
- **Foundation-mounted Panels**: For this type, the panels are mounted on concrete, driven piles, or helical piles.
- **Floating Panels**: The panels are mounted on a polymer structure that floats on a body of water. The water can also be an artificial basin or a lake.

2.9 Solar Panel Tilt

Solar panels are tilted to face with maximum solar radiation. There are few types of solar panel tilting as follows.

Fixed Tilt: For this type, the panel is tilted at a specific azimuth.

Multi-orientation Tilt: The panel can be tilted up to eight different orientations. For each orientation, there is a sub-array of panels. The multi-orientation tilt replaces the old "Heterogeneous fields" option.

Seasonal Tilt: The panel is tilted with two orientations, one for summer and one for winter.

Unlimited-shed Tilt: This tilting is used when sheds are too long with respect to their width.

Unlimited sun-shield Tilt: Same remarks as for unlimited-shed tilt apply for this type of tilting. The unlimited sun-shield tilting is suitable only for the south façades.

2.10 Solar Tracking System

The solar tracking system is a device that moves solar panels continuously to face the sun with the aim of maximizing the panels' output power. The tracking system is composed of three components: sun tracker, driving motor, and controller. It must be designed to withstand harsh weather conditions and to have a life of at least equal to that of the panels. The sun tracker is the part that follows the direction of sun precisely. The tracking signal from sun tracker inputs the controller, which commands the driving motor to move panels in a correct direction at specific orientation so as the panels are faced directly to the sun.

Solar tracking systems can be classified into one-axis and two-axis trackers as follows [13].

2.10.1 One-Axis Tracker

The one-axis tracking system is the best solution for small PV power plants. It has a single structure and less number of moving parts and therefore is less expensive than other tracking systems. The one-axis tracking system employs a tilted mount and one electric motor to move the panels on a trajectory relative to the Sun's position. The rotation axis can be horizontal, vertical, or oblique. Figure 2.5, shows a typical scheme of a one-axis tracker showing both the rotation axis (vector e) and the collector plane (vector normal to the collector plane). The angle between the two vectors is usually kept constant in this type of tracker.

There are several types of one-axis tracking system, which are explained below (Figure 2.6) [13].

2.10.1.1 North–South Horizontal-Axis Tracking

In this type, the tracker axis is horizontal and its direction is North–South and $X = 90°$.

2.10.1.2 Polar Tracking

North–South polar axis tilted on an angle equal to the latitude of the site. The rotation is adjusted in such a way that the tracker follows the meridian of the earth containing the sun. The angular velocity is 15°/h. For this type, the solar tracking system can be used both in the Northern latitudes and in places near the equator.

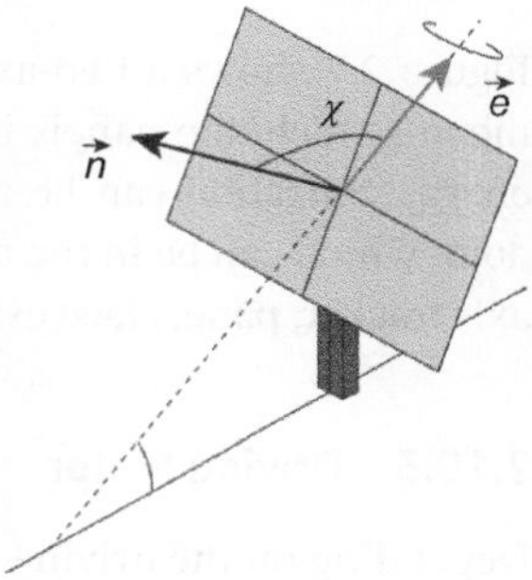
χ
$\vec{e}$
$\vec{n}$

Figure 2.5 Characteristic vectors in a one-axis tracker. *Source:* Reca-Cardeña and López-Luque [13].

2.10.1.3 East–West Horizontal-Axis Tracking

In this solar tracker type, the rotation axis is placed parallel to the ground and in an east–west direction. The panels can rotate only to follow the sun at

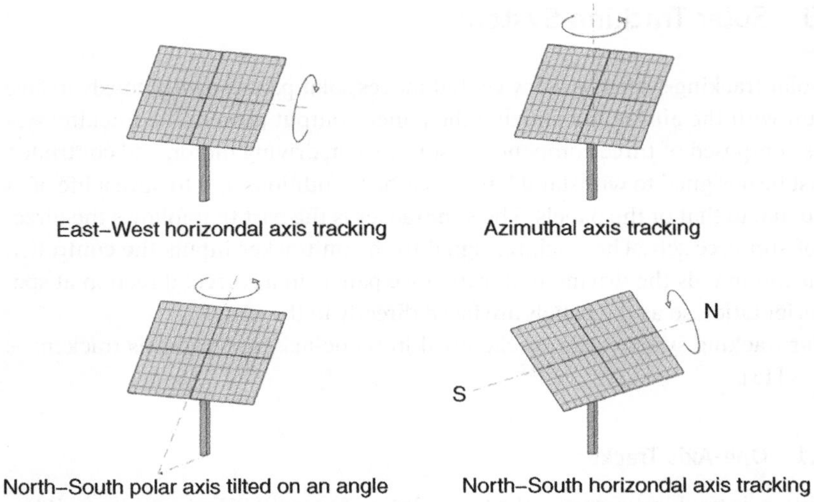
East–West horizondal axis tracking
Azimuthal axis tracking
N
S
North–South polar axis tilted on an angle
North–South horizondal axis tracking

Figure 2.6 Types of one-axis solar tracking systems.

its altitude angle and correct their position every day based on the Sun's declination. This tracker type is not commonly used because the produced panels' power is much lower than that of other types.

2.10.1.4 Azimuthal-Axis Tracking

The tilting angle of panels is constant and equal to the latitude of the site. The simple and robust mechanism of the azimuthal tracker with respect to the two axes trackers justifies their smaller collection of sun radiation, making them the most widely used tracker in practice.

2.10.2 Two-Axis Tracker

Figure 2.7 shows a two-axis solar tracking system. It provides the advantage of moving freely the panels in all directions. Using the two-axis tracker, maximum energy collection can be achieved due to the complete tracker movement freedom, which can be in the north–south and/or east–west directions. With the two-axis tracker, panels face exactly the sun's rays throughout the whole day.

2.10.3 Driving Motor

Depending on the driving mechanism, solar trackers can be classified into active or passive trackers. In the active tracker, the driving is performed by an electric DC/AC motor or a hydraulic system. In the passive tracker, the driving is carried

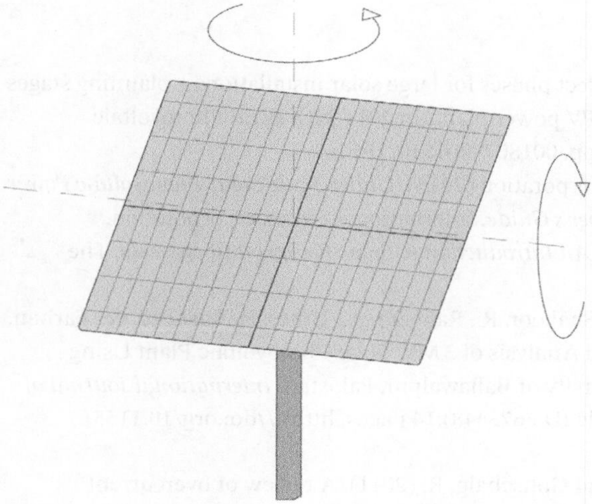

Figure 2.7 Two-axis solar tracking system.

out based on a gravitational system. The most common trackers are driven by an electric motor because it allows a simpler and precise control of the panel's movement. AC electric motors are preferred to the DC motors. For the one-axis tracker, only one motor is required, whereas for the two-axis tracker, two motors are needed [13].

2.10.4 Solar Tracker Control

Solar trackers can be classified in terms of their control into open-loop and closed-loop tracking systems. In the open-loop tracker, solar panels are oriented based on previously computed sun trajectories. The open-loop tracker does not require a sensor to detect the sun position. Instead, the sun movement is predicted based on astronomic relationships that are programmed in a microprocessor. The sun position is calculated by the microprocessor at any moment. The open-loop tracker is not affected by cloudiness or low-radiation circumstances that may lead to inaccurate tracking.

In the closed-loop tracker, panels are oriented based on live position of the sun, which is measured by a solar radiation sensor. The sensor is composed of photosensitive elements mounted on the panel. At dawn or under low-radiation conditions of cloudy days a tracker controlled by photo sensors may become disoriented. When disoriented, an auxiliary tracker controls the rotation of panels until the main tracker is restored.

References

1 Herfurth, D. (2011). Project phases for large solar installations – planning stages of Germany 5th largest PV power plant. In 2011 37th IEEE Photovoltaic Specialists Conference, pp. 001807–001810. IEEE.

2 International Finance Corporation (2015). *Utility-Scale Solar Photovoltaic Power Plants, A Project Developer's Guide*. International Finance Corporation.

3 J. Paul Guyer, P.E., R.A. *An Introduction to Solar Photovoltaic Systems*. The Clubhouse Press.

4 Faizan Ul Hassan Faiz, Shakoor, R., Raheem, A., Umer, F., Rasheed, N., Farhan, M. (2021). Modeling and Analysis of 3 MW Solar Photovoltaic Plant Using PVSyst at Islamia University of Bahawalpur, Pakistan. *International Journal of Photoenergy* 2021 (Article ID 6673448):14 pages. https://doi.org/10.1155/2021/6673448

5 Goss, B., Reading, C., and Gottschalg, R. (2011). A review of overcurrent protection methods for solar photovoltaic DC circuits. In: *5th Photovoltaic Science Application and Technology (PVSAT-5) Conference and Exhibition*. Edinburgh, Scotland: PVSAT.

6 Shertukde, H.M. (2014). *Distributed Photovoltaic Grid Transformers*, chapter 2 and p. 19. CRC Press.

7 Haider, R. and Kim, C.-H. (2016). Production of DERs. In: *Integration of Distributed Energy Resources in Power Systems* (ed. T. Funabashi), 157–192. Elsevier Academic Press.

8 Soga, T. (ed.) (2006). *Nanostructured Materials for Solar Energy Conversion*. Elsevier.

9 Global Solar Central Inverters Market by Type (Grid, Off-grid), by Application (Utilit, Non-utility) and by Region (North America, Latin America, Europe, Asia Pacific and Middle East & Africa), Forecast To 2028

10 Čorba, Z.J., Katić, V.A., Dumnić, B.P., and Milićević, D.M. (2012). In-grid solar-to-electrical energy conversion system modeling and testing. *Thermal Science* 16 (suppl. 1): 159–171.

11 Cabrera-Tobar, A., Bullich-Massagué, E., Aragüés-Peñalba, M., and Gomis-Bellmunt, O. (2016). Topologies for large scale photovoltaic power plants. *Renewable and Sustainable Energy Reviews* 59: 309–319.

12 Urishov, D. (2019). Yield modelling and validation of bifacial photovoltaic (PV) models for fixed tilted and single axis horizontal tracking systems for various climates in the world. Master thesis. Lappeenranta University of Technology, Finland.

13 Reca-Cardeña, J. and López-Luque, R. (2018). Design principles of photovoltaic irrigation systems. In: *Advances in Renewable Energies and Power Technologies* (ed. I. Yahyaoui), 295–333. Elsevier.

3

Feasibility Studies

3.1 Introduction

Feasibility studies are performed before the construction of a photovoltaic (PV) power plant. In these studies, the PV plant is evaluated from technical, legal, and economic points of view in order to analyze the issues such as the required space, shading of arrays, and access to the power grid. In this chapter, first, the key points and general definitions of feasibility studies of PV power plants are presented. Then, the criteria and requirements for feasibility studies report are presented.

3.2 Preliminary Feasibility Studies

Feasibility studies for large-scale PV power plants (LS-PVPPs) include two stages: preliminary feasibility studies and feasibility studies. In the preliminary studies, the feasibility of constructing a PV power plant is evaluated from technical, economic, and legal points of view. In these studies, the best site for the plant construction is selected based on various criteria. Moreover, obtaining permits is examined, and costs of constructing the PV plant are estimated. Once the initial plan for the PV plant is justified, the technical and economic feasibility studies are performed in more detail by experienced specialists and consultants.

A feasibility studies report is then prepared in accordance with the international standard. The report provides data related to the plant site including the solar radiation and meteorological information. Various feasibility studies to be conducted for a large-scale PV plant are shown in Figure 3.1. In the following sections, these studies are described.

Step-by-Step Design of Large-Scale Photovoltaic Power Plants, First Edition. Davood Naghaviha, Hassan Nikkhajoei, and Houshang Karimi.

3.3 Technical Feasibility Study

Technical feasibility study is related to the physical development of a PV plant. In the technical feasibility study, criteria related to the PV plant site selection are assessed.

3.3.1 Site Selection

Choosing a proper site for a LS-PVPP is greatly important because of its direct impact on the plant power generation, revenue, environment, and society. The following sections provide the most important site selection criteria.

3.3.1.1 Amount of Sunlight

One of the criteria in selecting the site is the amounts of daily and yearly solar radiation. In order to build a PV plant whose production revenue are commensurate with the plant capital, the amount of daily and annual solar radiations at the selected site must be estimated. The information needed to estimate the amount of solar radiation is available from global Atlases that are available on the national and international organizations websites.

3.3.1.2 Land Area and Geometry

The area required for each kilowatt-hour of a PV power plant depends on several factors including efficiency of the solar modules, amount of solar radiation, geographical location and geometric shape of the site land, shading distances, and the panel structure that can be fixed or movable.

The site should have adequate space to minimize shading distance, which reduces the PV production. In many countries, space is limited and the existing lands may be square or form multiple corners. For a country located in the Northern (Southern) hemisphere, the main direction of a land may not be toward the South (North). Therefore, when locating the solar arrays, a lot of empty spaces are created in the site location map.

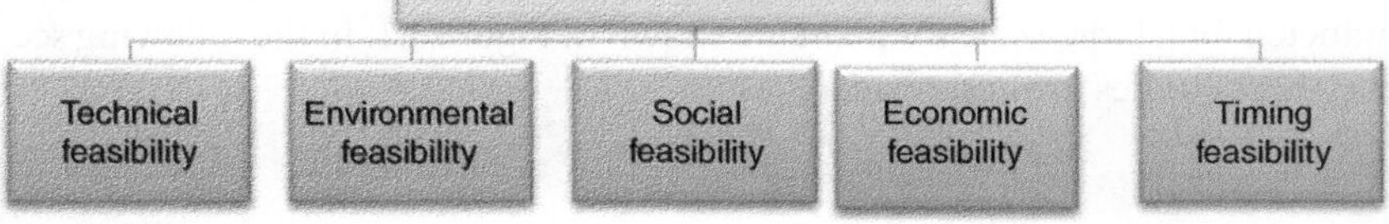
Feasibility studies
Technical feasibility
Environmental feasibility
Social feasibility
Economic feasibility
Timing feasibility

Figure 3.1 Feasibility study topics.

As a first priority, a land with a suitable shape should be selected in order to generate maximum power. As an example, five lands with different geometric shapes near the town of Twentynine Palms in the Southern San Bernardino, California, are shown in Figure 3.2. All lands have an area of about 30 ha. Moreover, the same type of solar modules in terms of efficiency and capacity have been installed. Land No. 1 is a square facing south with more solar panel capacity. As a result, the power obtained from Land No. 1 is the highest among the others.

The land area of a power plant can be estimated from the following equation.

$$A = \left(\frac{WP}{G \times \eta_M} \right) \times L_F \tag{3.1}$$

where A is the land area, WP denotes the total output power, η_M is the solar module efficiency, G is the solar irradiance, and L_F is the land factor, which is between 2 and 6 in the Central Europe. It should be noted that, in addition to the area calculated from (3.1), more land is needed for a PV plant, e.g. land corners, space required for the panels, and the other plant equipment.

3.3.1.3 Climate Conditions

It is greatly important to study the climate conditions of the region where the study site is selected. Climatic conditions affect the annual power production of a PV plant and the plant equipment. This, in turn, affects the costs of design and construction of PV power plant. For a study site, long-term meteorological data are collected to assess the climate conditions. If there is no meteorological station near the site, the required data is extracted using satellite meteorological software.

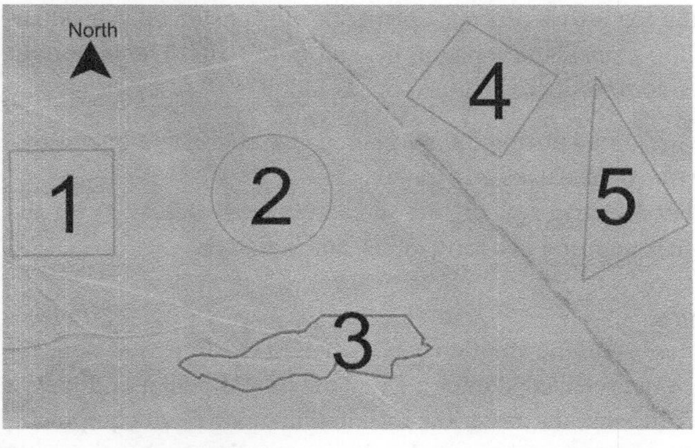
North
1
2
3
4
5

Figure 3.2 Five lands with different geometric shapes near the town of Twentynine Palms in southern San Bernardino, California.

For example, Table 3.1 shows the data obtained from Meteonorm satellite meteorological software for the city of Setúbal, Portugal.

The most important meteorological data necessary to evaluate a PV power plant site is wind speed and direction, ambient temperature, rainfall, humidity, and clouds scattering throughout the year. To calculate the mechanical forces of wind and snow on the structure supporting solar panels, these information must be collected for the site. In the areas where the amount of rainfall and cloud density is high, choosing solar panels with a proper technology has a great impact on the amount power produced by the PV plant.

3.3.1.4 Site Access to Power Grid

In feasibility studies, the access of PV plant site to substations and transmission lines, considering the followings, should be investigated.

- Site access to the power grid
- Privacy of the grid
- Proximity of the site to the grid
- Permission to connect the PV plant to the grid
- Legal permissions to construct substations and power transmission towers

The capacity of a PV plant cannot be more than the allowed capacity of the grid. The longer the distance from the PV plant site to the point of connection to the grid, the higher the costs associated with electricity transmission. In addition, the amount of power loss increases over a long transmission line, which delays the plant investment return.

3.3.1.5 Site Road Access

In the site selection, the roads to access a PV plant site are considered by evaluating the aerial or satellite maps, land topography, and field visits. The important points to be considered for the site road access are as follows:

- Access road for the personnel during the PV plant construction and operation
- Access road for the transport trucks and trailers
- Cultural, legal, or natural barriers along the site access road such as rocks, rivers, forests, cemeteries, monuments, sand fields, and wetlands

3.3.1.6 Site Topography

Site topography provides useful information such as height shadings and slope for a PV plant site. Ideally, the selected site for a PV plant should be flat and have a slight slope. The flatter the site, the easier the installation of PV plant equipment, and the lower the costs of construction, e.g. leveling, pounding, excavation, and land preparation.

Table 3.1 Satellite meteorological data from a station 30 miles south of the Portuguese capital in Setúbal near the north coast of the Sado River.

Month	G_Bn	Ta	UVA	el	Snd	D-RR	RR	FF	p	RH	N	Ts	G_Gh	G_Dh[1]
Jan	163	10.9	5	19.7	0	32	79	3.3	1006	81	4	10.4	101	38
Feb	157	12.1	7	24.5	0	28	55	3.4	1006	77	4	11.7	130	57
Mar	203	14.2	10	30	0	63	70	3.9	1006	72	4	14.6	185	68
Apr	240	15.4	14	35.8	0	52	57	4.1	1006	70	4	16.1	243	89
May	290	18.1	17	38.6	0	24	23	4	1007	67	3	19.4	291	96
Jun	320	21.6	19	39.5	0	6	14	4.1	1006	64	3	22.9	321	94
Jul	356	22.7	19	39.4	0	1	2	4.4	1006	62	3	24.3	327	79
Aug	338	23.3	17	37.7	0	0	5	4.2	1007	61	2	24.5	292	71
Sep	251	21.6	13	32	0	35	29	3.4	1007	65	4	22.1	225	72
Oct	186	18.6	9	26.8	0	61	95	3.7	1006	73	4	18.5	157	61
Nov	147	14.1	6	21.3	0	57	93	3.5	1006	77	5	13.4	108	46
Dec	147	11.6	4	19	0	79	96	3.7	1006	81	4	11	86	34
Year	234	17	12	31.6	0	438	618	3.8	1006	71	4	17.4	206	67

[1] G_Bn: Irradiance of beam; Ta: Air temperature; el: Height of sun; UVA: UVA radiation band; Snd: Snow depth; D-RR: Driving rain; RR: Precipitation; FF: Wind speed; p: Air pressure; RH: Relative humidity; N: Cloud cover fraction; Ts: Surface temperature; G_Gh: Mean irradiance of global radiation horizontal; G_Dh: Mean irradiance of diffuse radiation horizontal.

It is preferred that, in the Northern (Southern) hemisphere, the site should not be blocked by mountains or tall hills in the south (north). Depending on the distance of such obstacles to the PV site, the annual power production of the PV plant decreases.

A topography map with contour lines shows physical properties of the earth's surface on large and small scales. Figure 3.3 shows the topography map of a 200 000-ha area. The contour with higher density points are marked with darker colors which show higher altitudes. The central areas in Figure 3.3 show a plain away from high hills and mountains, which has a slight slope and is suitable for a PV power plant.

3.3.1.7 Land Geotechnics and Seismicity

A PV plant may be subject to earthquakes during the plant long-term operation. Therefore, it is necessary to conduct geotechnical studies before selecting a PV plant site. Geotechnical studies may include earthquake risk assessment,

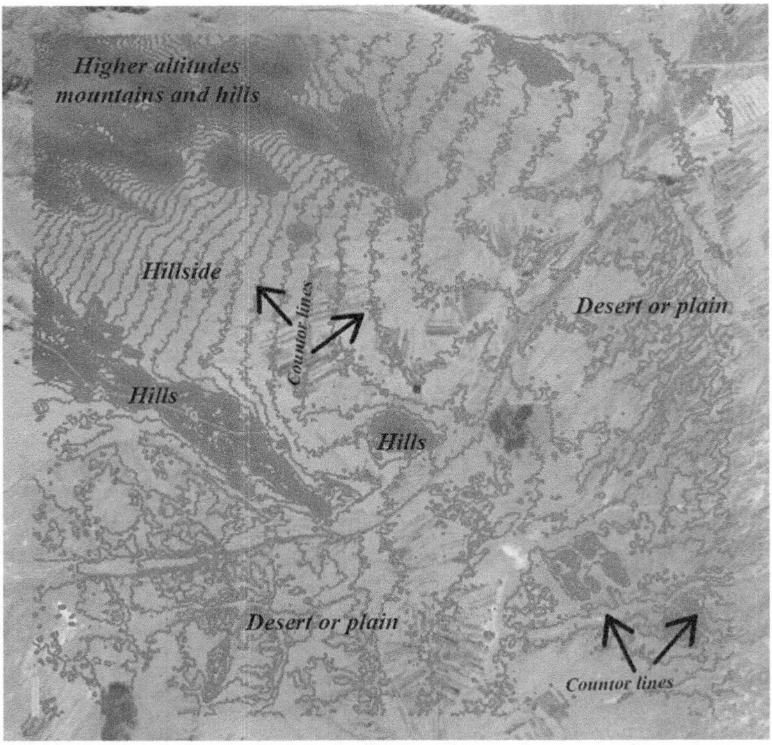
Higher altitudes
mountains and hills
Hillside
Countor lines
Desert or plain
Hills
Hills
Desert or plain
Countor lines

Figure 3.3 The topography map of a 200 000-ha area.

landslides, land subsidence, adjacent mining activities, soil susceptibility to frost, and soil erosion. The results of these studies are used for the following purposes:

- Design of foundation and structure bases
- Estimation of construction costs
- Estimation of construction life

For example, the soil test results in terms of the soil pH and chemical composition are used to evaluate the corrosion degree and to protect the plant structure against it by the use of a proper type of cement in the foundation concrete.

To conduct geotechnical studies, a number of pits with a depth of about 2.5–3 m are dug at certain intervals on the site ground. The soil samples of pits are collected and tested under the supervision of a geotechnical specialist in a laboratory. The site soil tests provide the following information:

- Groundwater level
- Soil resistance
- Presence of rocks and hard materials in the soil
- Soil permeability
- Soil compressive, tensile, and lateral strength
- Soil physical, mechanical, and chemical properties

3.3.1.8 Drainage, Seasonal Flooding

In addition to respecting the privacy of rivers, streams, swamps, and natural ponds, the PV plant site should not be selected in the path of canals and waterways. By examining satellite images and seasonal flood maps, its risk can be reduced to some extent. In case the selected site is subject to these risks, a dam, dyke or canal should be built to prevent any possible damages.

Figure 3.4 shows the path of seasonal floods in an area in which a 500-ha green site has been selected for the construction of a PV power plant. The area is completely dry and has a low average rainfall. However, by examining the seasonal flood map, it is determined that three seasonal floods pass through the site. Since the PV plants should operate over a long period, e.g. 30 years, they may be exposed to seasonal flooding and erosion of foundations and bases. Existing and new drainage should also be considered to ensure run-off is controlled to minimize erosion.

3.3.1.9 Land Use and Legal Permits

In the feasibility studies, the land use and legal permits of a PV plant site should be investigated by consulting the relevant organizations. Since PV power plants are usually built on low-value lands, the feasibility studies for land acquisition or rent, obtaining legal permits, and land transfer times should also be accounted.

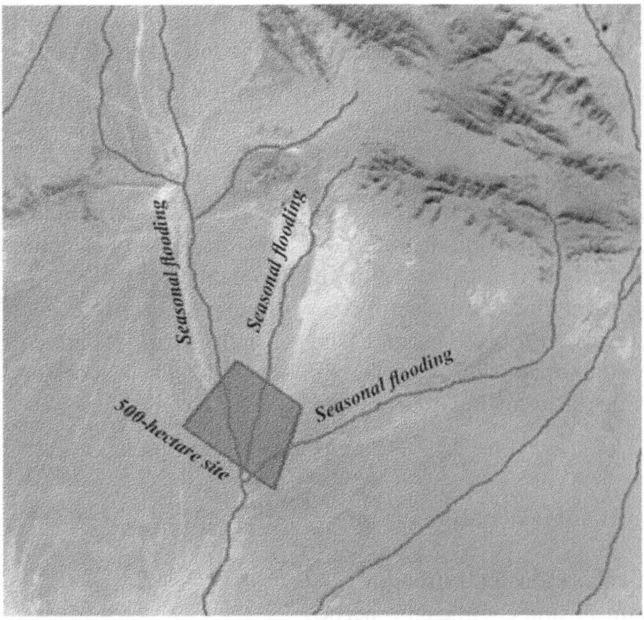
Seasonal flooding
Seasonal flooding
Seasonal flooding
500-hectare site

Figure 3.4 Satellite image of seasonal floods.

3.3.1.10 Air Pollution and Suspended Solid Particles

Collection of air pollution and quality indicators is necessary in order to estimate the actual power production of a PV plant. Air pollution is produced by the presence of pollutants in the air such as gases, vapors, dust, odors, or smoke. Air pollution significantly reduces the efficiency of PV panels. In addition, sulfur and other pollutants in the air can corrode PV plant equipment. If the plant is close to the sea, the level of salt in the atmosphere should also be examined. Salt can cause rapid corrosion of unprotected plant equipment. PV modules located in highly corrosive environments such as coastal areas must comply with the IEC 61701 standard for salt-dust corrosion.

Solid particles in the air include dust from desert areas or farmland, bird droppings, snow, and fine dust. The amounts of solid particles in the air including fine dust and solid pollution can be accurately determined by the soiling measurement method. In addition to reducing sunlight, airborne solids make the surface of solar panels dirty. Therefore, methods of cleaning the surface of PV modules should be considered. For example, in water-scarce areas, cleaning robots are used without the need for any water. In areas where there is more snow or air pollution, the modules should be installed at a greater angle of inclination so that the snow melts faster and the air pollution is washed away naturally by rainwater.

3.3.1.11 Geopolitical Risk

In choosing the site, political risks should be evaluated and predicted as they can significantly affect the profitability of PV power plants. Such policies include guaranteed electricity purchase rates, import and customs costs, and taxes.

3.3.1.12 Financial Incentives

PV power plants generally have high fixed costs. Construction of a PV plant leads to job creation and the expansion of clean energy use. For these reasons, there are generally incentives in many countries to attract investors to build PV plants. These include financial incentives such as high purchase rate of produced power, tax breaks, and customs exemptions.

3.3.2 Annual Electricity Production

The most important parameter after the site selection is the estimated annual power production of a PV plant. The amount of electricity generated by a plant is used to calculate its annual revenue and financial studies. Accurate estimation of the annual power production of a PV plant is carried out by specific software and simulations.

3.3.3 Equipment Technical Specifications

The technical specifications of a PV plant equipment must be compatible with the geographical and atmospheric conditions of the plant site, such as altitude, humidity, temperature, and other climatic parameters. In the feasibility studies, technical specifications of mechanical and electrical equipment are reviewed to ensure that a PV plant does not face significant breakdowns and consequences during its operation period.

3.3.4 Execution and Construction Processes

For the selected site and prior to the plant design, a roadmap for the execution and construction processes should be specified. The road map includes the execution and construction steps, and the way used for the plant construction and equipment installation. Drawing a roadmap ensures that no major risk or event during the PV plant construction is encountered.

3.3.5 Site Plan

The site plan of a PV plant shows the layout of solar panels and other equipment, taking into account the shading distances. The site plan specifies the land usable space, site geometry, land slope, proximity, access passages, climatic conditions, site topography, surface waterways, building dimensions, near and far shading distances, fences, switches and transformers, cable ducting, and plant output feeder.

Figure 3.5 shows the initial site plan of a PV power plant. The figure shows the shading distances, internal block of the power plant, locations of the inverters, fences, panels, warehouse, output feeder and the auxiliary cabin, and internal access roads.

3.4 Environmental Feasibility

PV power plants require large areas of land. Semidesert, desert, and industrial areas are suitable for PV power plants. Green areas and animal biodiversity areas should not be selected for PV plant construction. In the environmental feasibility studies, the following are considered:

- Environmental issues and their control methods
- Ways to compensate for damages to the environment after the plant implementation

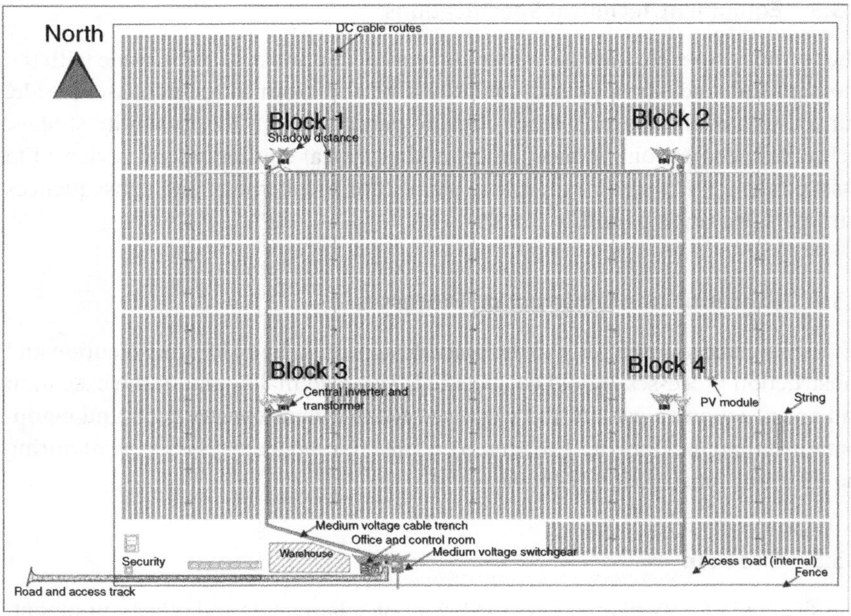
North
DC cable routes
Block 1
Shadow distance
Block 2
Block 3
Central inverter and transformer
Block 4
PV module
String
Medium voltage cable trench
Office and control room
Medium voltage switchgear
Warehouse
Security
Access road (internal)
Fence
Road and access track

Figure 3.5 Initial site plan of the solar power plant at a selected site.

- Methods of continuous monitoring and control of the plant construction and operation environmental impacts
- Extent of changes in plant and animal ecosystems and migration
- Rate of air pollution reduction
- Methods to recycle or dispose PV plant waste

3.5 Social Feasibility

Social feasibility studies are examined from two aspects that are explained below.

Human Resources: The availability of human resources needed for the construction and operation of a PV power plant should be studied. In addition, the possibility of providing the living needs of employees and workers should be considered. The living needs include accommodation, food, transportation, and access to health care, entertainment, social activities, telecommunications, and the Internet.

Social Justification: The social justification of constructing a PV power plant in addressing the needs of the neighboring society should be studied. If a PV plant does not meet the needs of its society, it is not socially justified. Social justification studies include direct and indirect employment, economic growth, impacts on the quality of life and health, population, and immigration.

3.6 Economic Feasibility

In economic feasibility studies, the financial model of a PV power plant is required in order to predict the investment and operating costs, and its revenues. By evaluating financial and economic indicators, the rate of return on investment can be optimized. The following sections describe the inputs of financial studies and the resulting financial indicators.

3.6.1 Financial Model Inputs

The inputs for financial model of a PV plant include its technical information and the sale price of produced power. The financial model inputs have seven main parts, each includes several components. Figure 3.6 shows the inputs to the financial model.

A) **Project Definition:** This includes basic PV plant information such as title, description and type (design or development), nominal capacity, construction and operation periods, the currency rates, and investment funds.

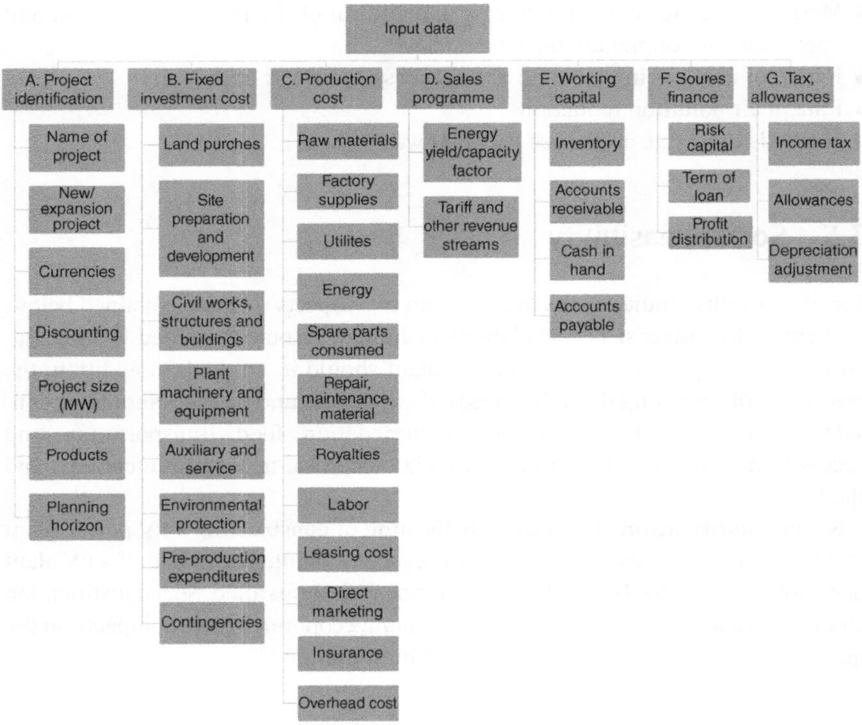
Input data
A. Project identification
Name of project
New/ expansion project
Currencies
Discounting
Project size (MW)
Products
Planning horizon
B. Fixed investment cost
Land purches
Site preparation and development
Civil works, structures and buildings
Plant machinery and equipment
Auxiliary and service
Environmental protection
Pre-production expenditures
Contingencies
C. Production cost
Raw materials
Factory supplies
Utilites
Energy
Spare parts consumed
Repair, maintenance, material
Royalties
Labor
Leasing cost
Direct marketing
Insurance
Overhead cost
D. Sales programme
Energy yield/capacity factor
Tariff and other revenue streams
E. Working capital
Inventory
Accounts receivable
Cash in hand
Accounts payable
F. Soures finance
Risk capital
Term of loan
Profit distribution
G. Tax, allowances
Income tax
Allowances
Depreciation adjustment

Figure 3.6 Important input parameters for financial modeling. *Source:* Adapted from Mahmoudi and Mahdavi [1].

B) **Investment Cost or Capex:** The investment cost is the sum of fixed costs of a PV power plant before starting the operation. The fixed costs includes the costs of land and construction of the plant buildings, the costs of installing the plant equipment, and the costs of obtaining the necessary permits, consulting, and training.

C) **Production Cost or Opex:** The production cost includes the costs of raw materials, employees, maintenance and administration, and etc.

D) **Electricity Sales Plan:** This includes the amount and price of electricity sales per year in order to estimate the PV plant's revenue. In the sales plan, the factors impacting on the plant power production are considered. These factors include the amount of sunlight, number of cloudy days in a year, ambient temperature, solar panel technologies and structure type (fixed or with tracker), the losses of cables and transformers, air pollution and dust particles, far and near shading distances, panels aging rate, guaranteed electricity purchase rate, adjustment rates, and government incentives.

E) **Current Capital:** This capital includes the total inventory of goods and cash, and the receivables including accounts payable.
F) **Financing Sources:** These sources include shareholder contributions, bank loans, and government subsidies.
G) **Tax:** This includes tax amount, exemptions, and reserves.

3.6.2 Financial Model Results

Based on the input data, the financial modeling and calculations for a PV plant are performed, and the results determine the financial performance of the plant for the investor. Figure 3.7 shows the results of financial model of an example PV power plant. The financial model results include the following:

A) **Estimated Investment Costs:** This includes estimates of Capex costs and the current capital required by the PV plant.
B) **Estimated Annual Production Costs:** This includes an estimate of the annual Opex costs required by the PV plant to cover the personnel, maintenance, and administrative costs.
C) **Revenue Analysis:** The analysis includes forecasts of revenues from the electricity sale during the PV plant operation.
D) **Sources of Financing:** This includes available financial resources to finance the PV plant and pay debts. These resources include investor funding, bank facilities, and bonds.

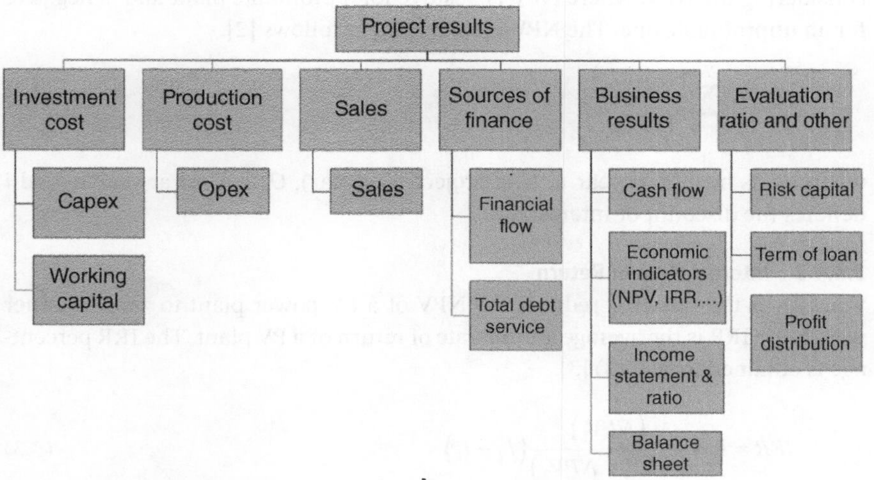
Project results
Investment cost
Production cost
Sales
Sources of finance
Business results
Evaluation ratio and other
Capex
Working capital
Opex
Sales
Financial flow
Total debt service
Cash flow
Economic indicators (NPV, IRR,...)
Income statement & ratio
Balance sheet
Risk capital
Term of loan
Profit distribution

Figure 3.7 The results of a financial model of an example of PV power plant.

E) **Investment Outcomes:** The investment outcome provides information about the cash flow of capital and economic indicators of the PV plant. The economic indicators include Internal Rate of Return (IRR), Net Present Value (NPV), Break-Even Point (BEP), profitability index, sensitivity, PI risk indices, and profit and loss.
F) **Financial Performance Indices:** This shows the results of analyzing financial ratios and performance indices of the PV power plant.

3.6.3 Financial and Economic Indicators

The construction of a PV plant is justified if it has sufficient revenue to pay the debts and general expenses of Operations and Maintenance (O&M). Moreover, the PV plant should have a reasonable profit. To financially evaluate the PV plant, first one must carefully estimate its fixed costs, power production, and the revenue-obtained from electricity sales. For a PV power plant, if financial indices such as the IRR, NPV, and payback period are appropriate, the construction of plant is justifiable and profitable.

3.6.4 Financial Indicators

For a PV power plant, the most important financial indicators are explained below.

3.6.4.1 Net Present Value

The NPV is the difference between present values of the input and the output cash flows. The amount of capital required to build a PV power plant is calculated by considering the NPV, where NPV is positive for a profitable plant and is negative for an unprofitable one. The NPV is calculated as follows [2].

$$NPV = \sum_{n=0}^{N} \frac{(B_n - C_n)}{(1+i)^n} \tag{3.2}$$

where B_n is benefit at year n, N is project life (year), C_n is Cost at year n, and i denotes the discount or interest rate.

3.6.4.2 Internal Rate of Return

The IRR is the rate that reduces the NPV of a PV power plant to zero. In other words, the IRR is the average annual rate of return of a PV plant. The IRR percentage is obtained from (3.3) [3].

$$IRR = I_1 + \frac{(NPV_1)}{(NPV_1 - NPV_2)}(I_2 - I_1) \tag{3.3}$$

where I is the discount or interest rate.

The *IRR* can also be calculated from

$$IRR = \sum_{t-1}^{t} \frac{(B_n - C_n)}{(1 + IRR)^n} - C_0 = 0 \tag{3.4}$$

where

B_n = Benefit at year n
C_n = Cost at year n
C_0 = initial investment

3.6.4.3 Investment Return Period

The investment return period is the period during which the capital for construction of a PV power plant is returned, taking into account the time value of money and the rate of capital loss. The investment return period is calculated as follows [3, 4].

$$PP = \sum_{n=0}^{PP} \frac{(B_n - C_n)}{(1 + i)^n} = 0 \tag{3.5}$$

where

B_n = Benefit at year n
C_n = Cost at year n
i = Discount or interest rate

3.6.4.4 Break Even Point

BEP is a condition in which revenues and expenses of a PV power plant are equal. In other words, the revenue obtained from the produced electricity compensates the fixed and variable costs of power production, and thus the net profit of the PV plant is zero (Figure 3.8). The BEP is determined as follows [5].

Figure 3.8 shows that a PV power plant is profitable and economically justified when its costs are to the right of the BEP point.

$$BEP = \frac{F}{(S - V)} \tag{3.6}$$

BEP = Break-Even Point
F = Total fixed costs
V = Variable costs per unit of power production
S = Savings or additional returns per unit of power production

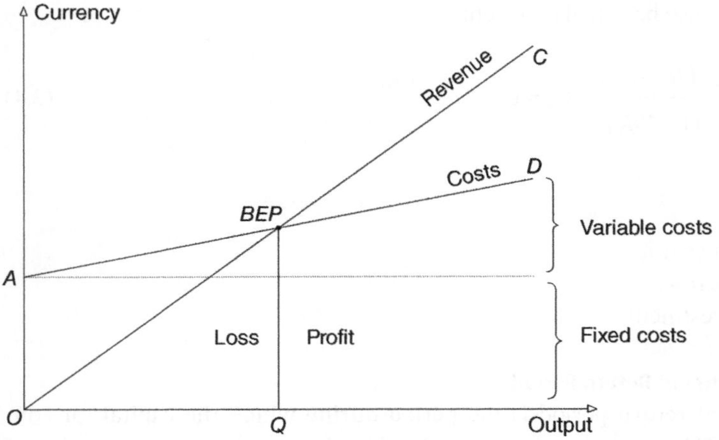
Currency
Revenue
C
Costs
D
BEP
Variable costs
A
Loss
Profit
Fixed costs
O
Q
Output

Figure 3.8 A plot showing break-even point.

3.7 Timing Feasibility

Scheduling and control is one of the most important aspects of managing the construction of a PV power plant. In feasibility studies, completion time of the PV power plant is estimated. The shorter the construction time, the higher the IRR value of the PV power plant will be. The timing feasibility studies include the following:

- PV power plant phase schedule
- Work breakdown structure (WBS) and activities
- Time of each activity
- Gantt chart of activities
- Funding resources
- Plant construction scheduling and duration, cost, and executive resources
- Critical path of the PV plant and time–cost relationship

Using the results of time feasibility studies, the problems that may occur during construction or operation of a PV power plant can be predicted and prevented.

3.8 Summary

The owner or builder of a PV power plant must have a thorough knowledge of the technical issues and local regulations related to the plant site before obtaining legal permits and constructing the PV plant. This knowledge is obtained through feasibility studies in five fields including technical, economic, social, environmental, and timing. The feasibility of a PV power plant is evaluated in order to prevent major problems and financial damages during the PV plant construction and operation.

References

1 Mahmoudi, A. and Mahdavi, M. (2011). Application of an international standard pattern for financial and economical evaluation of the tourism services projects (case study Rijab-Dalahou City-Kermanshah Province). *Indian Journal of Science and Technology* 4 (6): 708–715.

2 Hsieh, S.C. (2014). Economic evaluation of the hybrid enhancing scheme with DSTATCOM and active power curtailment for PV penetration in Taipower distribution systems. *IEEE Transactions on Industry Applications* 51 (3): 1953–1961.

3 Haramaini, Q., Setiawan, A., Damar, A. et al. (2019). Economic analysis of PV distributed generation investment based on optimum capacity for power losses reducing. *Energy Procedia* 156: 122–127.

4 Menéndez, J. and Loredo, J. (2019). Economic feasibility of developing large scale solar photovoltaic power plants in Spain. In: *E3S Web of Conferences*, vol. 122, 02004. EDP Sciences.

5 Tsorakidis, N., Papadoulos, S., Zerres, M., and Zerres, C. (2011). *Break-even Analysis*. Bookboon.

4

Grid Connection Studies

4.1 Introduction

Connecting distributed generation sources such as photovoltaic (PV) power plants to the power grid affects its operation, stability, and safety. The harmful effects include injection of current harmonics into the power network, increase in voltage at the connection point, voltage and flicker fluctuations, increase in fault currents, and voltage and frequency instability. Hence, before connecting a PV power plant to the grid, it is necessary to study and evaluate the technical issues and standards of grid connection. In case of noncompliance with the standards, ancillary equipment should be used to reduce adverse impacts of the PV power plant.

Technical studies of the grid connection of a PV power plant are performed using an advanced simulation software based on the national network codes and standards. The technical studies include load flow, short circuit, power quality, and stability analysis. In the simulation software, electrical feasibility of connecting the PV plant to the grid is evaluated.

4.2 Introducing Topics of Grid Connection Studies

The main topics of PV plant grid connection studies are shown in Figure 4.1. In addition, costs and time of connection to the grid, and transmission line and network capacities and the corresponding legal and technical constraints should be assessed.

4.2.1 Load Flow Studies

Load flow studies are performed for the cases of presence and absence of PV power plant in different load conditions. The objectives of load flow studies are to

Step-by-Step Design of Large-Scale Photovoltaic Power Plants, First Edition. Davood Naghaviha, Hassan Nikkhajoei, and Houshang Karimi.

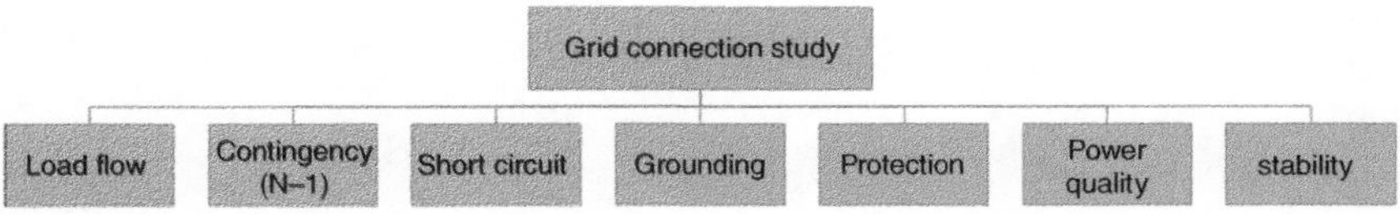
Grid connection study
Load flow
Contingency (N–1)
Short circuit
Grounding
Protection
Power quality
stability

Figure 4.1 Block diagram representation of grid connection study report.

evaluate voltage profiles of different buses, power network losses, and loads on transformers and lines. In load flow studies, the following considerations should be taken into account:

- In the power network model, the network arrangement, amounts of loads, and equipment electrical parameters before and after the PV plant connection should be included.
- Load flow studies should be performed at peak load and low load. If the amount of losses in low load increases when the PV plant is connected to the grid, the average load condition should also be studied. Based on the results of average load losses the effect of PV plant on the network losses is assessed.
- Peak load and low load values should be related to the day times when the PV power plant produces power. In cases where the PV plant has an energy storage system, the peak and low load values should be related to a 24-hour duration. In this case, the peak and low loads do not necessarily coincide with the maximum and minimum of PV plant power production [1].
- To obtain a more accurate result for load flow studies, the effect of PV plant can be evaluated in a long-term period. For this purpose, the regional load growth and the reduction coefficients of PV plant power production in coming years are considered in the load flow studies.
- Detailed load flow studies in the presence of PV power plant should be performed based on the production curve and the grid load curve for a specified period, preferably one year. With a detailed load flow study, full impacts of the PV power plant on the grid parameters can be well determined.

4.2.2 Contingency (N-1)

The purpose of contingency studies is to determine the power network conditions when removing one of the main equipments of the network. Often the worst case scenario is when a key equipment is disconnected during peak load conditions. In emergency studies, the results of load flow studies should be considered for few emergency scenarios. When the PV plant is connected to the medium voltage bus of a distribution substation, the scenarios of N-1 conditions are:

- Loss of a substation transformer;
- Loss of a substation feeder;
- Disconnection of lines between the PV power plant and a substation; and
- Loss of an upstream transmission transformer.

The most important indicators that are examined in the emergency studies are the state of voltage amplitude in different network buses and the amount of equipment load. Note that the acceptable ranges for voltage and equipment load in emergency situations are wider than their amounts under normal conditions. For example, the grid operation instructions may accept a voltage range of up to ±10% per unit for emergencies, while this range is ±5% for normal load conditions.

4.2.3 Three-phase and Single-phase Short Circuit Studies

The purpose of short circuit studies is to investigate thermal effect of fault currents on power network components, coordination of protection equipment and relays, and determination of the cutoff power required for power switches. For the short circuit studies, the following cases should be investigated:

- Various operating scenarios
- Presence and absence of the PV power plant
- Occurrence of various fault types
- Faults at different locations in the downstream and upstream of the PV power plant
- Faults inside the PV power plant and in its adjacent feeders

The areas adjacent to the point of common coupling (PCC) are a priority for the short circuit studies since the connection of a PV power plant to the grid has the greatest impact on the PCC and its adjacent areas. For this reason, short circuit studies at the PCC and its adjacent areas are essential. In addition, the impacts of connecting a PV power plant to the grid on the required cutoff power of the switches at the PCC and the adjacent areas should be investigated. Short circuit studies should be performed for existing inverters in the PCC area to determine the effect of connecting the PV power plant to the grid. The capabilities of these inverters can be used to control short circuit currents. Inverters usually have the ability to participate in fault current in order to maintain their terminal voltage. Inverters can inject up to five times their rated current for a short period of a few tens of milliseconds [2].

4.2.4 Grounding System Studies

Having a safe and standard grounding system is essential for the PV power plants. The grounding system is one of the most important components for the proper functioning of PV power plant protection systems and personnel safety. The grounding system is designed to protect people and equipment against overvoltages generated in the body of the equipment to prevent contact and unwanted voltages. The grounding system must be designed in such a way as:

- to create a suitable path for the circulation of current through the ground in normal conditions and fault in order to prevent the large current passing through the equipment.
- not to be exposed to dangerous voltage in case the ground is connected to an equipment or a person.
- not to produce excess voltage in the area due to the passage of surface current gradient.
- to provide a small total resistance so that the rate of increase of earth voltage or ground potential rise (GPR) due to the passage of current is small.

In the grid connection studies of PV power plant, the maximum single-phase fault current to ground and the share of current passing through the ground network are calculated. Using grounding system analysis software and the IEEE-80-2000 standard, a ground network must be designed that meets the allowable limits for the contact and step voltages, and GPR and the ground resistance [3].

4.2.5 Network Protection Studies

Short circuit in a PV power plant can have many adverse effects. This includes high currents and damages to equipment of the PV power plant and the power network. The equipment that may be damaged includes inverters, transformers, cables, transmission lines, and switches. In addition, due to short circuits, the voltage of the power network buses drops sharply and adversely impacts quality of the power supplied to consumers and the stability of generators. Therefore, the PV power plant protection system should be designed properly as to provide high protection against short circuit. The protection system is designed based on the method of isolating a limited area around the fault location in the shortest time.

For PV power plants, the typical types of protection mentioned in Table 4.1 are considered. For LS-PVPP, special considerations for the plant protection should be taken into account. In a LS-PVPP, unlike the conventional radial-powered networks, power is fed to consumers from all sides. Therefore, the size, duration, and direction of fault currents are different from conventional networks. Thus, the fault current may become out of the range of network protection equipment and lead to a malfunction of the protection equipment [4].

In addition, LS-PVPP may reduce the grid current, resulting in delayed operation or failure of the grid protection equipment. In addition, depending on the connection type of PV power plant transformer, the inverters may change the

Table 4.1. The typical types of protection for PV power plants.

Protective functions	
81o/v: Frequency relay	Phase fault:
TT: Transfer trip	51V: Overcurrent relay
81R: Rate of change frequency relay	67: Directional Overcurrent
27/59: Under voltage relay/overvoltage relay	21: Distance relay
59I: Overcurrent relay	
25: Synchronism-check relay	32: Directional power relay
47: Phase sequence voltage relay	Grounding faults:
46: phase balance current relay	51N, 67N, 59N, 27N
78: out-of-step protective relay	

zero-sequence component current for a single-phase fault to ground and affect the performance characteristic of protection relays [4].

The controllability of production capacity of LS-PVPP during the day is greatly important for the network protection. According to international standards [5], due to the lack of inertia of PV power plants, their production capacity must be controlled in case of network emergency. The output power of a PV system depends on the weather conditions, where the cloudy weather leads to a rapid decrease in its output. This is important in LS-PVPP; since, with a sharp decrease in the output power, the network frequency exceeds its allowable limit. For a certain amount of radiation reduction, there is enough time for the governor system of synchronous generators to compensate for the production loss of the PV plant. Otherwise, under frequency load (UFL) reduction relays must cut off part of the network load to increase the frequency [6].

4.2.6 Power Quality Studies

Providing excellent power quality with minimum electricity interruption for consumers is greatly important. A large decline in the quality of network power causes damage to sensitive industries and consumers. Connecting LS-PVPP to the grid should not reduce the quality of grid power. In case of violation of the power quality, the power plant investor is obliged to use ancillary equipment or control methods to comply with the standards.

Any problem that causes a change in voltage, current, or frequency and thus leads to malfunction of consumer equipment is related to power quality. Major power quality disturbances include transient disturbances, outages, short-term voltage drops, short-term voltage surges, waveform distortions, voltage fluctuations, frequency changes, and flicker.

LS-PVPP can increase the voltage at the point of grid connection, especially at noon on summer days. In addition, changing the output power of the PV power plant due to changes in weather conditions can lead to voltage fluctuations [7]. In addition, inverters in a PV power plant can greatly increase the amount of harmonics injected to the power network. Studying the harmonics produced by inverters is more important than other power quality parameters. In most commercial inverters, the current THD is less than 3%. However, depending on the network background harmonics, harmonic injection with a 3% current THD may lead to an increase of harmonics beyond the standard range [8].

Solar inverters can provide reactive power at night. The network voltage status at night should be studied to evaluate the capability of PV plant to control the reactive power. Relevant standards should be considered for the control of reactive power by a PV power plant. If the reactive power of a LS-PVPP is not sufficient, reactive power compensation equipment such as a capacitive bank should be used. In such cases, the possibility of capacitive bank resonance with the network impedances should be investigated. If resonance occurs in the characteristic harmonics, it is necessary to consider a capacitive bank as a filter to remove the harmonics.

4.2.7 Stability Studies

Connecting a LS-PVPP to the power grid may cause instability, depending on the capacity of the PV plant, its installation location, the location and type of fault, and the operating time of the protection relays. In power systems, stability studies are investigated in three areas of rotor angle stability, including small signal stability and transient stability, frequency stability, and voltage stability.

LS-PVPP have the important capability to control real power for the frequency stability and reactive power for the voltage stability [9]. In terms of the rotor angle stability, connecting a PV power plant to the grid has a positive effect. However, in the case of small signal stability, the damping effect of PV power plant depends on the operating point of the system and can be positive or negative.

The real power of a large-scale PV power plant can improve frequency stability of a power network. The requirements for real power control of the PV power plant for frequency stability are provided in the international standards [6]. According to the international standards, it is necessary to conduct frequency stability studies for various operating conditions, including a sudden change in the PV plant output power, e.g. during the cloudy weather.

In terms of voltage stability, PV power plants should contribute by injecting reactive power to maintain voltage stability, e.g. during short circuit events. The requirements for reactive power control by the PV power plant are provided in the international standards [6].

The impacts of large-scale PV power plant on the performance of the grid protection system should be studied. Due to the variable nature of PV power plants, power swings in transmission lines may occur when the plant is connected to the grid. These fluctuations should be studied when examining the performance of the differential and directional relays of the transmission lines. On the other hand, power fluctuations with high amplitude that occur during the cloudy weather affect the stability of the other neighboring power systems [8]. Therefore, the impacts of the PV power plant on the stability of the network and adequacy of the protection system must be carefully studied.

4.3 Modeling of Grid and PV Power Plants

PV power plant grid connection studies are performed with power system simulation software, including DIgSILENT, Etap, Cyme, PSS-E, EMTP, and PSCAD. In this section, first the information required for the modeling of a PV power plant and the power network is described. Next, a sample PV power plant connected to the grid is modeled in DIgSILENT software. Then, a summary of the notions presented in Section 4.2 is provided.

4.3.1 Background Information Required for Modeling

The information required for modeling of a PV power plant and the power network in the DIgSILENT software are as follows:

- Technical specifications of PV plant components, e.g. PV modules, inverters, internal transformers, and power transmission lines;
- Location of the PV power plant and its distance from a public substation or feeder;
- Voltage ranges of distribution and transmission systems[1];

[1] Voltage ranges of the power systems in the United States are as follows:

1. The distribution system has voltages from 7 to 13 kV
2. The high voltage distribution system has voltages of 34, 46, and 69 kV
3. The transmission system is divided to three groups of high voltage (HV), extra high voltage (EHV), and ultra high voltage (UHV) systems. The voltages of HV system are 115, 138, 161, and 230 kV, and those of EHV system are 345, 500, and 765 kV. The UHV system has a voltage greater than 765 kV.

Voltage ranges of the power systems in Europe and Asia are as follows:

1. The distribution system has voltages from 10 to 33 kV
2. The high voltage distribution system has voltages of 63, 110, 132, and 150 kV
3. The transmission system has voltages of 230 and 400 kV.

- Allowable network frequencies;
- The grid parameters including symmetrical three-phase short circuit level, three-phase short-circuit current, X/R ratio of the network such as impedances Z1, Z2, X1, X0, and R0;
- Technical specifications of substation transformers including their capacity, conversion ratio, group, and X/R ratio;
- Single-line network diagram;
- The amount of network load in low, medium, and peak operating conditions;
- Annual load curve; and
- Annual electrical energy produced by the PV power plant.

Figure 4.2 shows a single-line diagram of a power grid which shows the information needed for modeling of the electrical components. The information required to model a PV power plant is shown in the single-line diagram in Figure 4.3. This information is related to the equipment, substation, and internal transmission line of the PV power plant. The information related to the PV power plant and the grid should be obtained by the owner or operator of the PV plant.

In Figure 4.3, multiple PV modules form a series of strings and parallel connection of the strings creates a PV block. The output DC voltage of each PV block is boosted up by a DC–DC converter which is then converted to AC voltage by a central inverter. The inverter output is connected to an internal transformer to increase the voltage from low to medium level. The most important information required for the PV power plant modeling is shown next to each section of the PV power plant in Figure 4.3.

4.3.2 Simulation of PV Plant and Network

After receiving the power network and PV plant information, the grid-connected PV plant is modeled using a network analysis software such as DIgSILENT. Figure 4.4 shows the complete network and PV plant diagrams in DIgSILENT software. The simulation diagram of Figure 4.4 is ready for the studies of load flow, short circuit, power quality, and dynamics. In the next sections, the results of various studies for grid-connected PV power plant are presented.

4.3.3 Load Flow Studies Before and After PV Plant Connection

The results of load flow studies for the cases of before and after PV plant connection are shown in Figures 4.5 and 4.6, respectively.

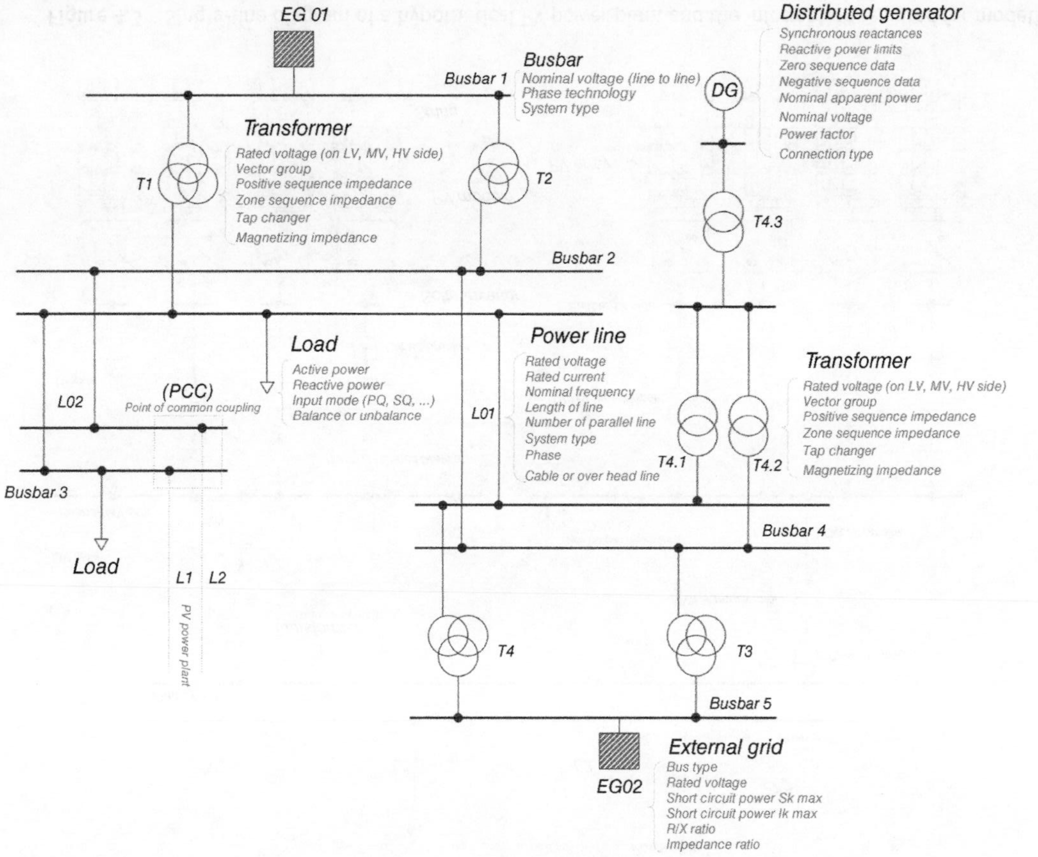
EG 01
Busbar
Busbar 1
Nominal voltage (line to line)
Phase technology
System type
DG
Distributed generator
Synchronous reactances
Reactive power limits
Zero sequence data
Negative sequence data
Nominal apparent power
Nominal voltage
Power factor
Connection type
Transformer
Rated voltage (on LV, MV, HV side)
Vector group
Positive sequence impedance
Zone sequence impedance
Tap changer
Magnetizing impedance
T1
T2
T4.3
Busbar 2
Load
Active power
Reactive power
Input mode (PQ, SQ, ...)
Balance or unbalance
Power line
Rated voltage
Rated current
Nominal frequency
Length of line
Number of parallel line
System type
Phase
Cable or over head line
Transformer
Rated voltage (on LV, MV, HV side)
Vector group
Positive sequence impedance
Zone sequence impedance
Tap changer
Magnetizing impedance
(PCC)
Point of common coupling
L02
L01
T4.1
T4.2
Busbar 3
Busbar 4
Load
L1
L2
PV power plant
T4
T3
Busbar 5
External grid
Bus type
Rated voltage
Short circuit power Sk max
Short circuit power Ik max
R/X ratio
Impedance ratio
EG02

Figure 4.2 Single-line diagram of a hypothetical electrical grid and the information required for modeling.

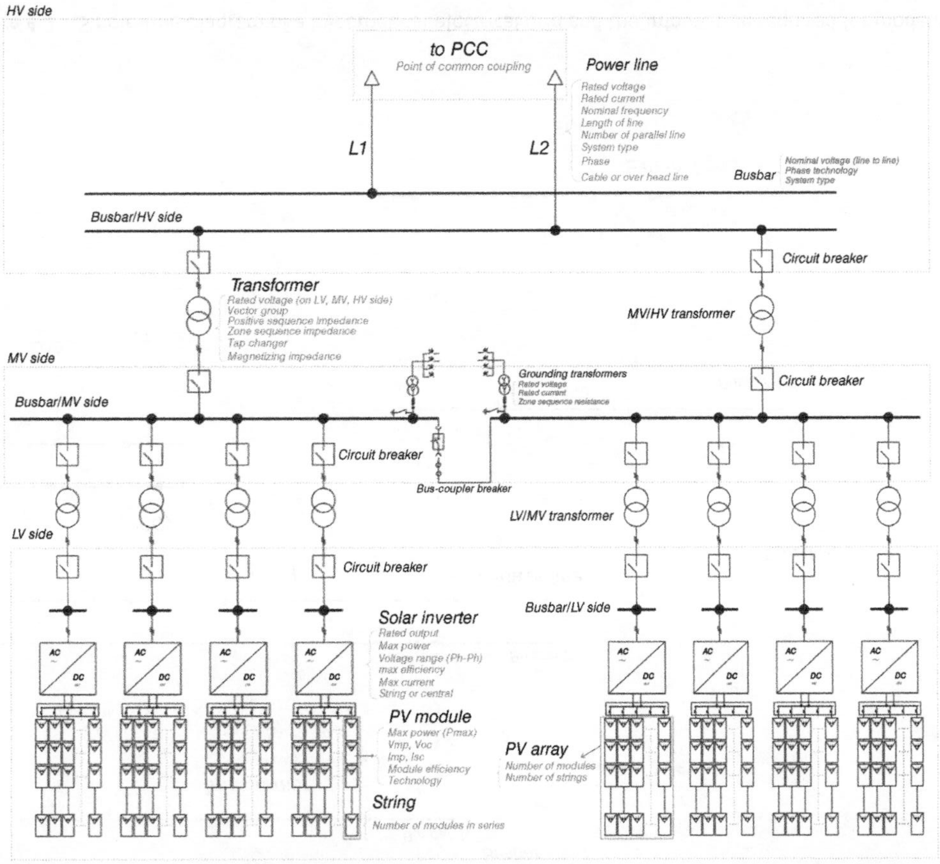
HV side
to PCC
Point of common coupling
Power line
Rated voltage
Rated current
Nominal frequency
Length of line
Number of parallel line
System type
Phase
Cable or over head line
L1
L2
Busbar
Nominal voltage (line to line)
Phase technology
System type
Busbar/HV side
Circuit breaker
Transformer
Rated voltage (on LV, MV, HV side)
Vector group
Positive sequence impedance
Zone sequence impedance
Tap changer
Magnetizing impedance
MV/HV transformer
MV side
Grounding transformers
Rated voltage
Rated current
Zone sequence resistance
Circuit breaker
Busbar/MV side
Circuit breaker
Bus-coupler breaker
LV/MV transformer
LV side
Circuit breaker
Busbar/LV side
Solar inverter
Rated output
Max power
Voltage range (Ph-Ph)
max efficiency
Max current
String or central
AC
DC
PV module
Max power (Pmax)
Vmp, Voc
Imp, Isc
Module efficiency
Technology
PV array
Number of modules
Number of strings
String
Number of modules in series

Figure 4.3 Single-line diagram of a hypothetical PV power plant and the information required for modeling.

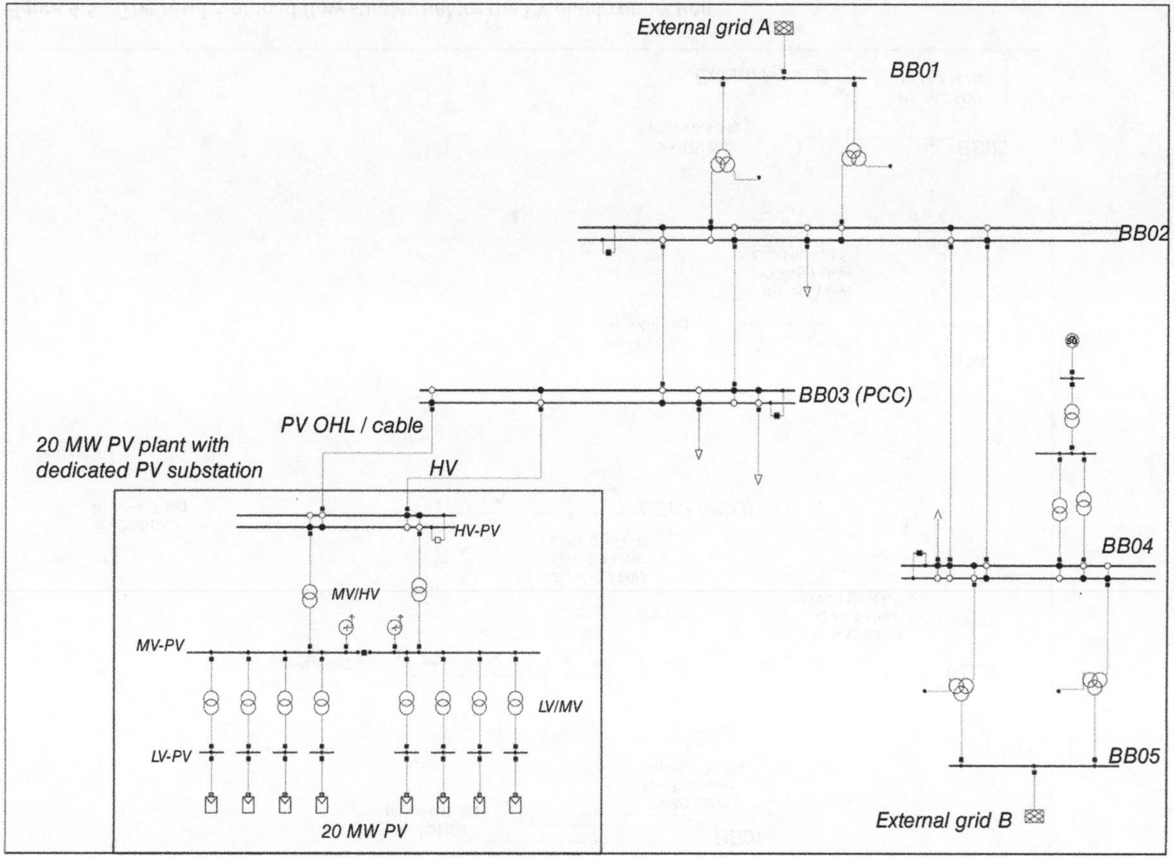
External grid A
BB01
BB02
BB03 (PCC)
SG
PV OHL / cable
20 MW PV plant with
dedicated PV substation
HV
HV-PV
BB04
MV/HV
MV-PV
LV/MV
LV-PV
BB05
20 MW PV
External grid B

Figure 4.4 The complete electrical network and PV plant diagrams in DIgSILENT software.

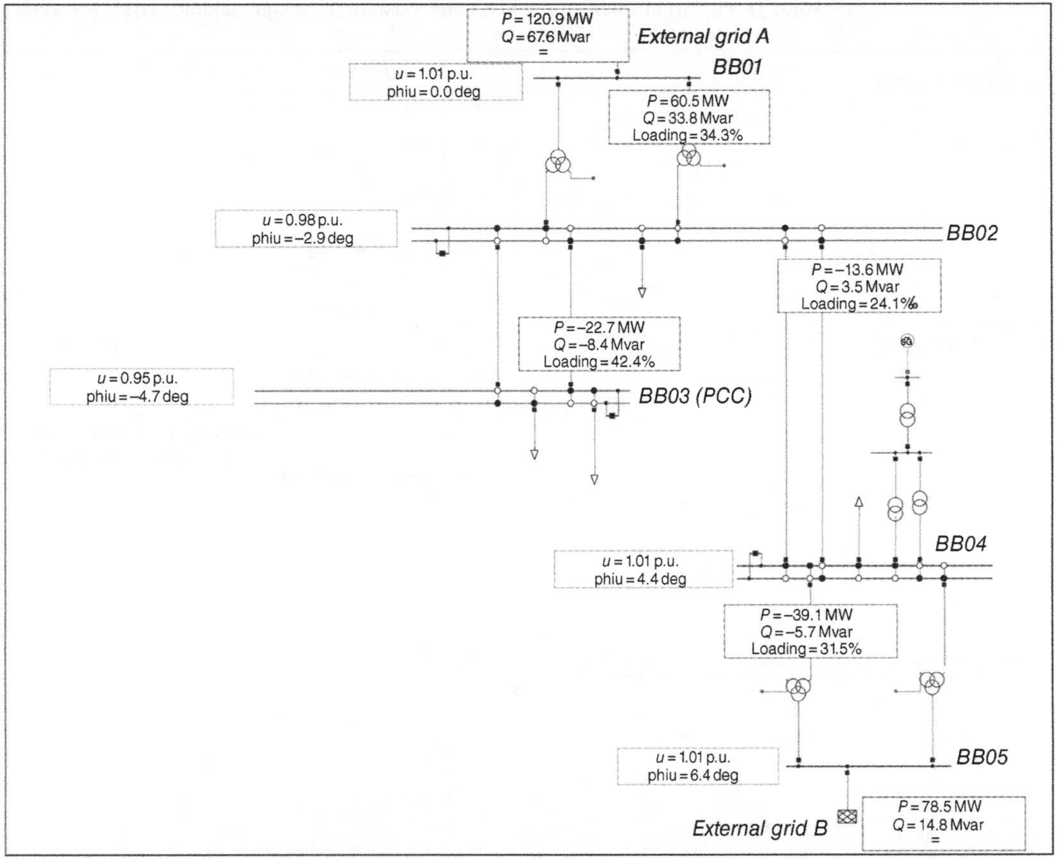
P = 120.9 MW
Q = 67.6 Mvar
=
External grid A
BB01
u = 1.01 p.u.
phiu = 0.0 deg
P = 60.5 MW
Q = 33.8 Mvar
Loading = 34.3%
u = 0.98 p.u.
phiu = −2.9 deg
BB02
P = −13.6 MW
Q = 3.5 Mvar
Loading = 24.1%
P = −22.7 MW
Q = −8.4 Mvar
Loading = 42.4%
u = 0.95 p.u.
phiu = −4.7 deg
BB03 (PCC)
BB04
u = 1.01 p.u.
phiu = 4.4 deg
P = −39.1 MW
Q = −5.7 Mvar
Loading = 31.5%
u = 1.01 p.u.
phiu = 6.4 deg
BB05
External grid B
P = 78.5 MW
Q = 14.8 Mvar
=

Figure 4.5 The results of load flow studies before the PV plant connection.

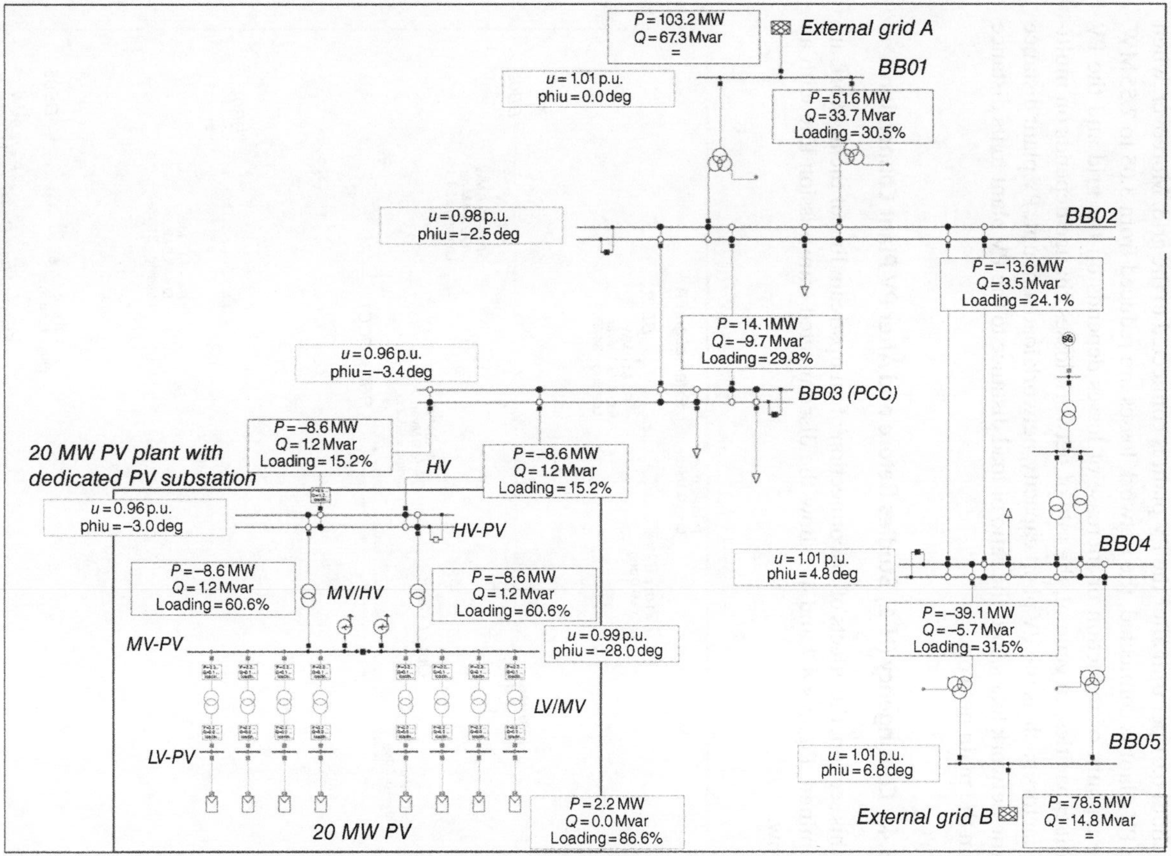
External grid A
P = 103.2 MW
Q = 67.3 Mvar
=
BB01
u = 1.01 p.u.
phiu = 0.0 deg
P = 51.6 MW
Q = 33.7 Mvar
Loading = 30.5%
BB02
u = 0.98 p.u.
phiu = −2.5 deg
P = −13.6 MW
Q = 3.5 Mvar
Loading = 24.1%
P = 14.1 MW
Q = −9.7 Mvar
Loading = 29.8%
u = 0.96 p.u.
phiu = −3.4 deg
BB03 (PCC)
20 MW PV plant with
dedicated PV substation
P = −8.6 MW
Q = 1.2 Mvar
Loading = 15.2%
HV
P = −8.6 MW
Q = 1.2 Mvar
Loading = 15.2%
u = 0.96 p.u.
phiu = −3.0 deg
HV-PV
P = −8.6 MW
Q = 1.2 Mvar
Loading = 60.6%
MV/HV
P = −8.6 MW
Q = 1.2 Mvar
Loading = 60.6%
MV-PV
u = 0.99 p.u.
phiu = −28.0 deg
LV/MV
LV-PV
20 MW PV
P = 2.2 MW
Q = 0.0 Mvar
Loading = 86.6%
BB04
u = 1.01 p.u.
phiu = 4.8 deg
P = −39.1 MW
Q = −5.7 Mvar
Loading = 31.5%
BB05
u = 1.01 p.u.
phiu = 6.8 deg
External grid B
P = 78.5 MW
Q = 14.8 Mvar
=

Figure 4.6 The results of load flow studies after the PV plant connection.

The results show that connection of the PV plant to the grid reduces the load of power transformers and transmission lines. The voltage range in the PCC improves by about 0.01 per unit after the PV plant is connected to the grid. Moreover, when the PV plant is connected, the network losses are reduced from 3.05 to 2.55 MW. The amount of reduction or increase of losses depends on the grid and the PV plant capacities. In general, the rate of network losses change depends on multiple factors such as the PV plant capacity, network load amount, PV plant distance to the network loads, and the ratio of load distance to the PV plant to its distance from the main network.

4.3.4 Contingency (N-1) Studies Before and After PV Plant Connection

In this section, the effects of disconnection of an upstream line of the network are examined. Figures 4.7 and 4.8 show the disconnected transmission line with an arrow.

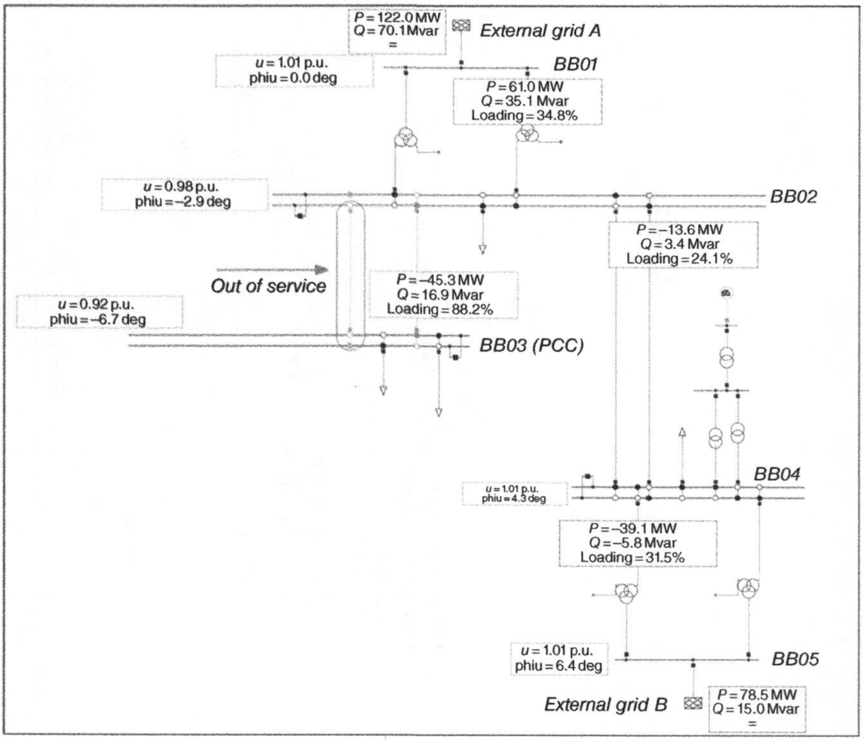
P = 122.0 MW
Q = 70.1 Mvar
External grid A
u = 1.01 p.u.
phiu = 0.0 deg
BB01
P = 61.0 MW
Q = 35.1 Mvar
Loading = 34.8%
u = 0.98 p.u.
phiu = –2.9 deg
BB02
P = –13.6 MW
Q = 3.4 Mvar
Loading = 24.1%
Out of service
P = –45.3 MW
Q = 16.9 Mvar
Loading = 88.2%
u = 0.92 p.u.
phiu = –6.7 deg
BB03 (PCC)
BB04
u = 1.01 p.u.
phiu = 4.3 deg
P = –39.1 MW
Q = –5.8 Mvar
Loading = 31.5%
u = 1.01 p.u.
phiu = 6.4 deg
BB05
External grid B
P = 78.5 MW
Q = 15.0 Mvar

Figure 4.7 Contingency (N − 1) studies before the PV plant connection.

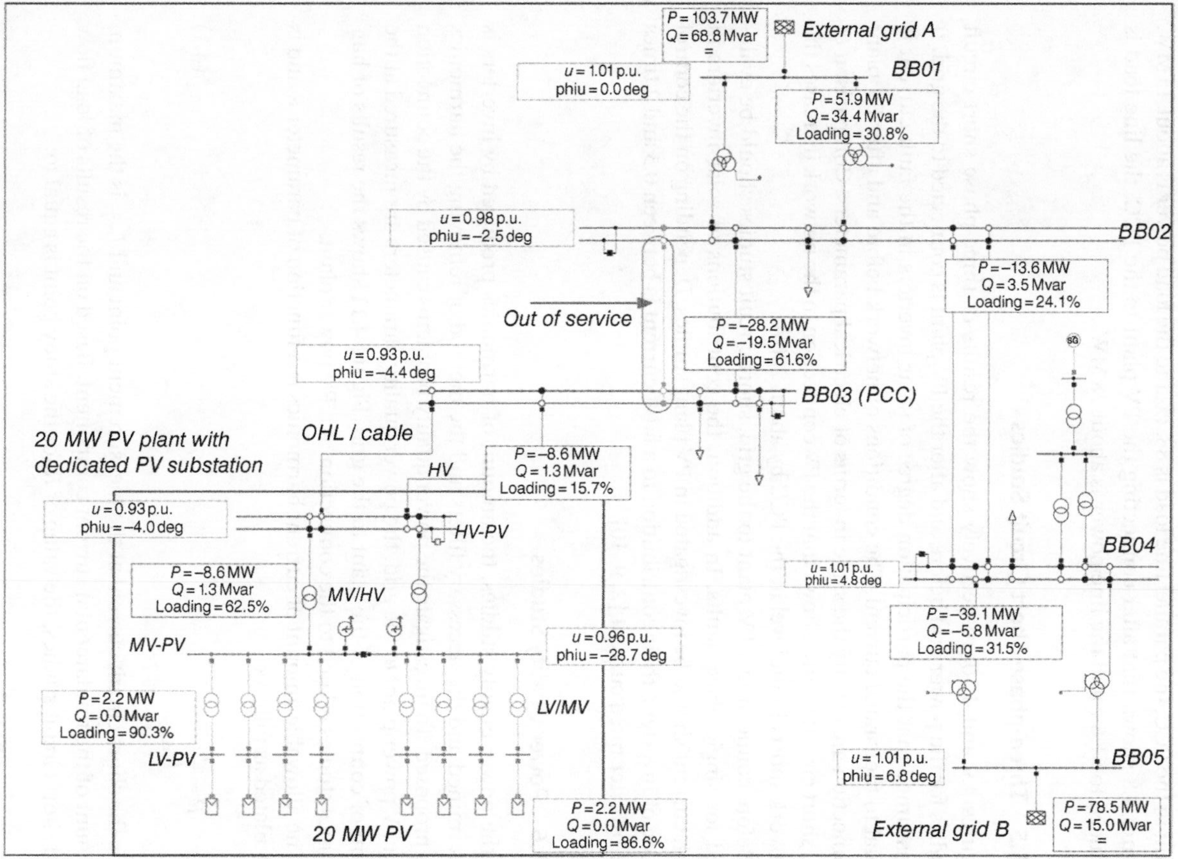
P = 103.7 MW
Q = 68.8 Mvar
External grid A
u = 1.01 p.u.
phiu = 0.0 deg
BB01
P = 51.9 MW
Q = 34.4 Mvar
Loading = 30.8%
u = 0.98 p.u.
phiu = −2.5 deg
BB02
P = −13.6 MW
Q = 3.5 Mvar
Loading = 24.1%
Out of service
P = −28.2 MW
Q = −19.5 Mvar
Loading = 61.6%
u = 0.93 p.u.
phiu = −4.4 deg
BB03 (PCC)
20 MW PV plant with
dedicated PV substation
OHL / cable
HV
P = −8.6 MW
Q = 1.3 Mvar
Loading = 15.7%
u = 0.93 p.u.
phiu = −4.0 deg
HV-PV
P = −8.6 MW
Q = 1.3 Mvar
Loading = 62.5%
MV/HV
MV-PV
u = 0.96 p.u.
phiu = −28.7 deg
P = 2.2 MW
Q = 0.0 Mvar
Loading = 90.3%
LV/MV
LV-PV
20 MW PV
P = 2.2 MW
Q = 0.0 Mvar
Loading = 86.6%
BB04
u = 1.01 p.u.
phiu = 4.8 deg
P = −39.1 MW
Q = −5.8 Mvar
Loading = 31.5%
u = 1.01 p.u.
phiu = 6.8 deg
BB05
External grid B
P = 78.5 MW
Q = 15.0 Mvar

Figure 4.8 Contingency (N − 1) studies after the PV plant connection.

The conditions in both cases, before and after the PV connection, are identical in terms of electrical parameters. Figure 4.7 shows that before connecting the PV plant to the PCC, the parallel line load is 88.2% and the load power is about 45 MW. Figure 4.8 shows that after connecting the PV plant to the PCC, the line load is reduced to 61.6% and the line power is about 28 MW.

4.3.5 Three-phase Short Circuit Studies

Figures 4.9 and 4.10, respectively show the results of three-phase short circuit studies for the power grid before and after the PV plant is connected to the grid. It is assumed that the participation degree of solar inverters in the fault current is equal to their rated current. The conditions of network before and after PV plant connection are exactly the same in terms of electrical parameters. Comparison of the short circuit results shows that the PV connection to the network increases the network short circuit level in the PCC by about 200 A.

Before connecting a PV plant to the grid, short circuit studies should be evaluated for single-phase faults. In addition, the contributions of solar inverters in fault current should be investigated in PV plant studies. Depending on the control mode of inverters, their contribution to a fault current is between 0.5 and 2 times the inverter rated current [2, 9, 10].

4.3.6 Power Quality Studies

In the power quality studies, the amount of harmonics produced by inverters is determined and the necessary filters and the method of removing the harmonics are proposed. To investigate the power quality problems caused by the connection of a PV power plant to the grid, the power quality parameters are measured at the point of connection of the plant to the grid. Figure 4.11 shows the results of harmonics studies related to the connection of a PV power plant.

The allowable amount of current harmonics is a function of parameter R and is calculated as follows.

$$R = \frac{I_{sc}}{I_{\max}} \tag{4.1}$$

I_{sc} is the short-circuit current at the measurement point and I_{max} is the maximum amount of the fundamental current component. Based on the results of load flow and short circuit studies, the value of R for the study point is equal to:

$$R = \frac{8200}{166} = 49.3 \tag{4.2}$$

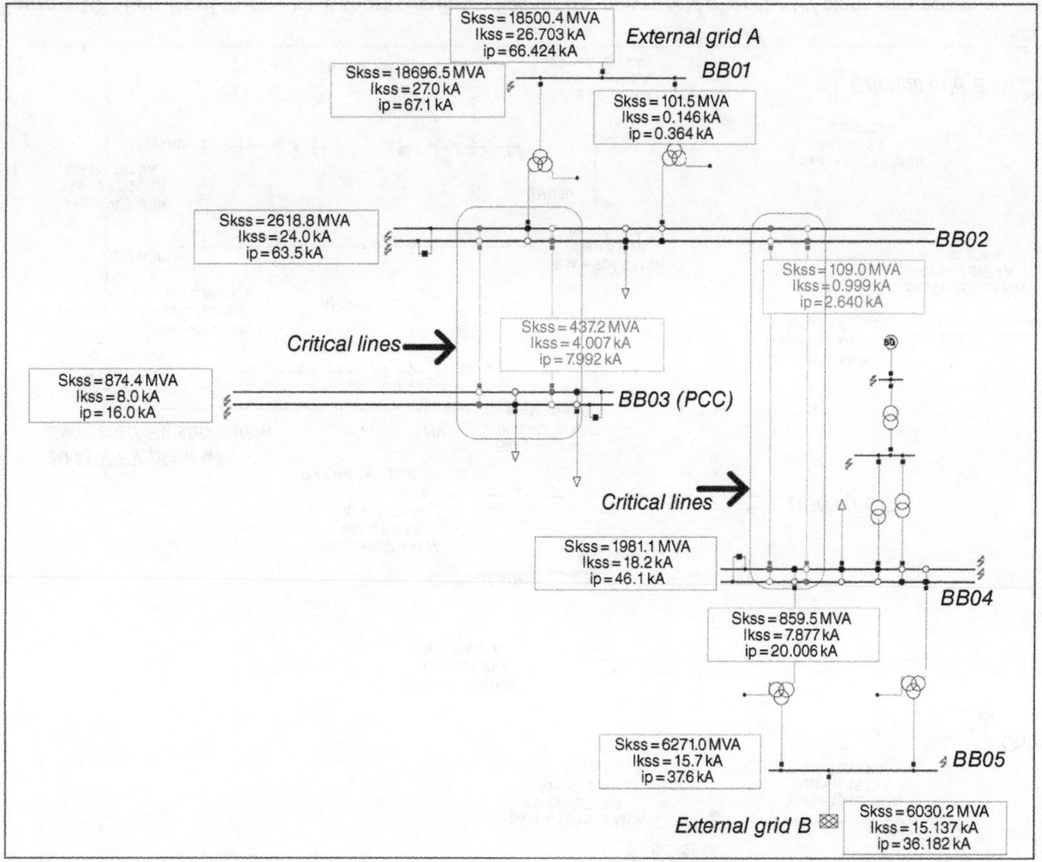

Skss = 18500.4 MVA
Ikss = 26.703 kA
ip = 66.424 kA
External grid A
Skss = 18696.5 MVA
Ikss = 27.0 kA
ip = 67.1 kA
BB01
Skss = 101.5 MVA
Ikss = 0.146 kA
ip = 0.364 kA
Skss = 2618.8 MVA
Ikss = 24.0 kA
ip = 63.5 kA
BB02
Skss = 109.0 MVA
Ikss = 0.999 kA
ip = 2.640 kA
Critical lines
Skss = 437.2 MVA
Ikss = 4.007 kA
ip = 7.992 kA
Skss = 874.4 MVA
Ikss = 8.0 kA
ip = 16.0 kA
BB03 (PCC)
Critical lines
Skss = 1981.1 MVA
Ikss = 18.2 kA
ip = 46.1 kA
BB04
Skss = 859.5 MVA
Ikss = 7.877 kA
ip = 20.006 kA
Skss = 6271.0 MVA
Ikss = 15.7 kA
ip = 37.6 kA
BB05
External grid B
Skss = 6030.2 MVA
Ikss = 15.137 kA
ip = 36.182 kA

Figure 4.9 Results of three-phase short circuit studies for the power grid before the PV plant connection.

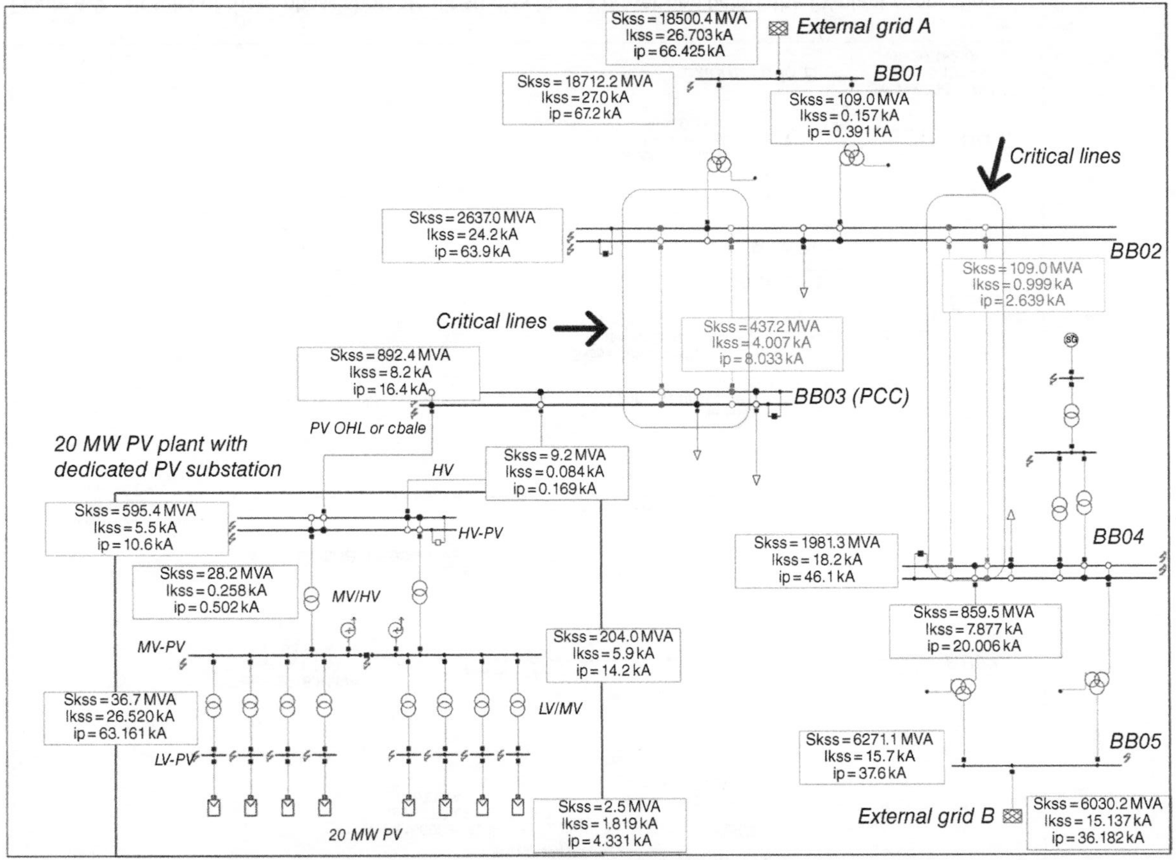
Skss = 18500.4 MVA
Ikss = 26.703 kA
ip = 66.425 kA
External grid A
BB01
Skss = 18712.2 MVA
Ikss = 27.0 kA
ip = 67.2 kA
Skss = 109.0 MVA
Ikss = 0.157 kA
ip = 0.391 kA
Critical lines
Skss = 2637.0 MVA
Ikss = 24.2 kA
ip = 63.9 kA
BB02
Skss = 109.0 MVA
Ikss = 0.999 kA
ip = 2.639 kA
Critical lines
Skss = 437.2 MVA
Ikss = 4.007 kA
ip = 8.033 kA
Skss = 892.4 MVA
Ikss = 8.2 kA
ip = 16.4 kA
BB03 (PCC)
PV OHL or cbale
20 MW PV plant with
dedicated PV substation
HV
Skss = 9.2 MVA
Ikss = 0.084 kA
ip = 0.169 kA
Skss = 595.4 MVA
Ikss = 5.5 kA
ip = 10.6 kA
HV-PV
Skss = 28.2 MVA
Ikss = 0.258 kA
ip = 0.502 kA
MV/HV
MV-PV
Skss = 204.0 MVA
Ikss = 5.9 kA
ip = 14.2 kA
Skss = 36.7 MVA
Ikss = 26.520 kA
ip = 63.161 kA
LV/MV
LV-PV
20 MW PV
Skss = 2.5 MVA
Ikss = 1.819 kA
ip = 4.331 kA
BB04
Skss = 1981.3 MVA
Ikss = 18.2 kA
ip = 46.1 kA
Skss = 859.5 MVA
Ikss = 7.877 kA
ip = 20.006 kA
BB05
Skss = 6271.1 MVA
Ikss = 15.7 kA
ip = 37.6 kA
External grid B
Skss = 6030.2 MVA
Ikss = 15.137 kA
ip = 36.182 kA

Figure 4.10 Results of three-phase short circuit studies for the power grid after the PV plant connection.

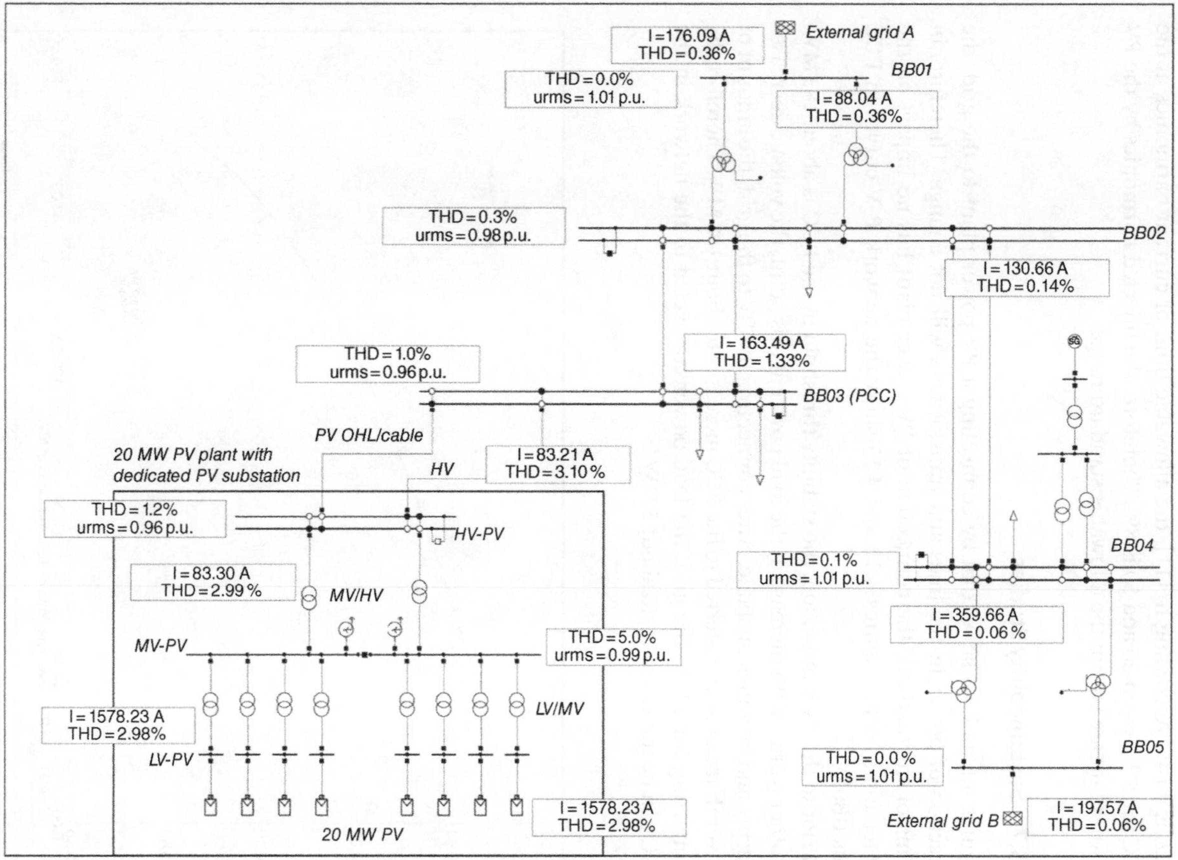
External grid A
I=176.09 A
THD=0.36%
BB01
THD=0.0%
urms=1.01 p.u.
I=88.04 A
THD=0.36%
THD=0.3%
urms=0.98 p.u.
BB02
I=130.66 A
THD=0.14%
I=163.49 A
THD=1.33%
THD=1.0%
urms=0.96 p.u.
BB03 (PCC)
PV OHL/cable
20 MW PV plant with
dedicated PV substation
HV
I=83.21 A
THD=3.10%
THD=1.2%
urms=0.96 p.u.
HV-PV
I=83.30 A
THD=2.99%
MV/HV
MV-PV
THD=5.0%
urms=0.99 p.u.
I=1578.23 A
THD=2.98%
LV/MV
LV-PV
I=1578.23 A
THD=2.98%
20 MW PV
BB04
THD=0.1%
urms=1.01 p.u.
I=359.66 A
THD=0.06%
BB05
THD=0.0%
urms=1.01 p.u.
External grid B
I=197.57 A
THD=0.06%

Figure 4.11 The results of harmonics studies related to the connection of a PV power plant.

The harmonic distortion rate at the PCC is 3.1%. With a nominal line current of 166 A and the short circuit current of 8200 A at the connection point, the I_{sc}/I_{load} ratio is 49.40. According to [11], the allowable limit of current harmonics at the PCC for R between 20 and 50 is 8%. Therefore, the injected harmonics by the PV power plant are within the allowable standard range.

4.3.7 Sustainability Studies

Figures 4.5 and 4.6 show that by connecting a PV power plant to the grid, the direction of power in the lines and transformers will not change. Therefore, in terms of fluctuations, the connection of PV power plant has no impact on the protection system. Figures 4.12 and 4.13 show the network P–V curves in PCC (bus BB03).

Prior to the PV connection, the stability threshold in the PCC is about 183 MW. In other words, with increase in the load power at this point, the voltage decreases sharply and becomes unstable. After connecting the PV to the grid, the amount of power that can be transferred to the PCC increases to about 188 MW. Therefore, by connecting the PV to the grid, the load power connected to the network in the PCC can be increased up to about 5 MW.

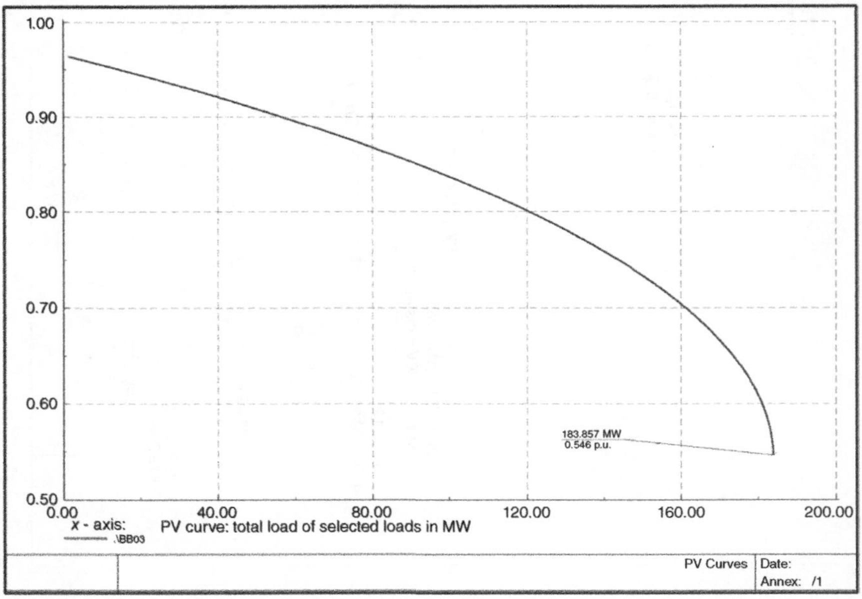
1.00
0.90
0.80
0.70
0.60
0.50
0.00
40.00
80.00
120.00
160.00
200.00
183.857 MW
0.546 p.u.
x - axis: PV curve: total load of selected loads in MW
\BB03
PV Curves
Date:
Annex: /1

Figure 4.12 Power curve before connecting the PV to the PCC.

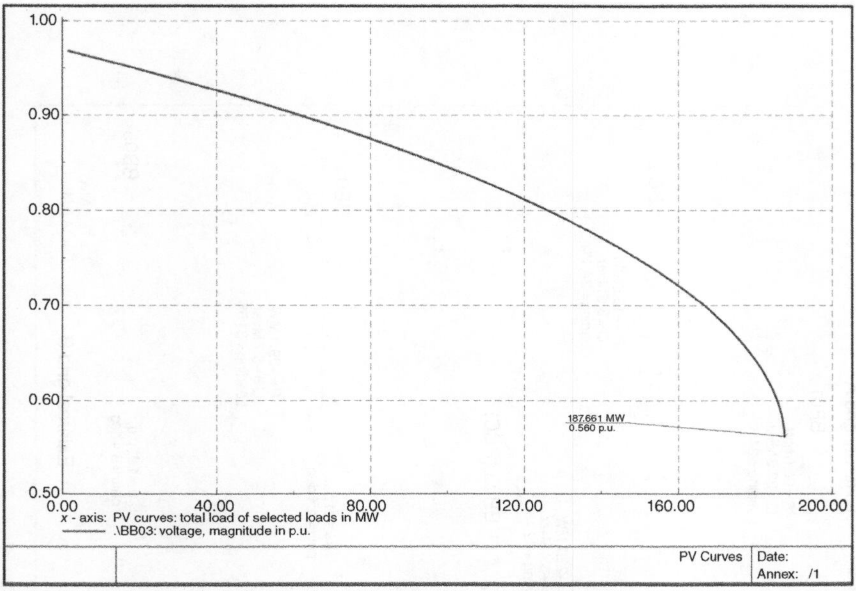
1.00
0.90
0.80
0.70
0.60
0.50
0.00
40.00
80.00
120.00
160.00
200.00
187.661 MW
0.560 p.u.
x - axis: PV curves: total load of selected loads in MW
.\BB03: voltage, magnitude in p.u.
PV Curves
Date:
Annex: /1

Figure 4.13 Power curve after connecting the PV to the PCC.

4.3.8 Investigating Additional Parameters for Grid Connection Studies

PV plant grid connection studies may indicate that the power network faces minor and/or major problems. These problems must be resolved to obtain an initial permit for PV connection. Then, it is necessary to verify the following:

- Costs of connecting the PV power plant to the grid
- Time of connecting the PV plant to the grid
- Capacities and technical and legal constraints of the grid and transmission line
- Transmission lines ohmic losses and transformers losses

The costs of purchasing equipment are required to estimate the fixed investment costs of the PV power plant. The time of connecting the PV power plant to the grid must be specified to estimate the execution time for construction of the dedicated distribution substation and transmission line. Estimation of the final PV plant construction time is required to prepare a financial studies report for the PV plant.

For the 20 MW PV power plant studied in the previous sections, the losses of various sections according to their technical specifications are shown in Figure 4.14. The network losses for the cases of before and after PV connection to the PCC are shown in Figures 4.14 and 4.15, respectively. The network loss before

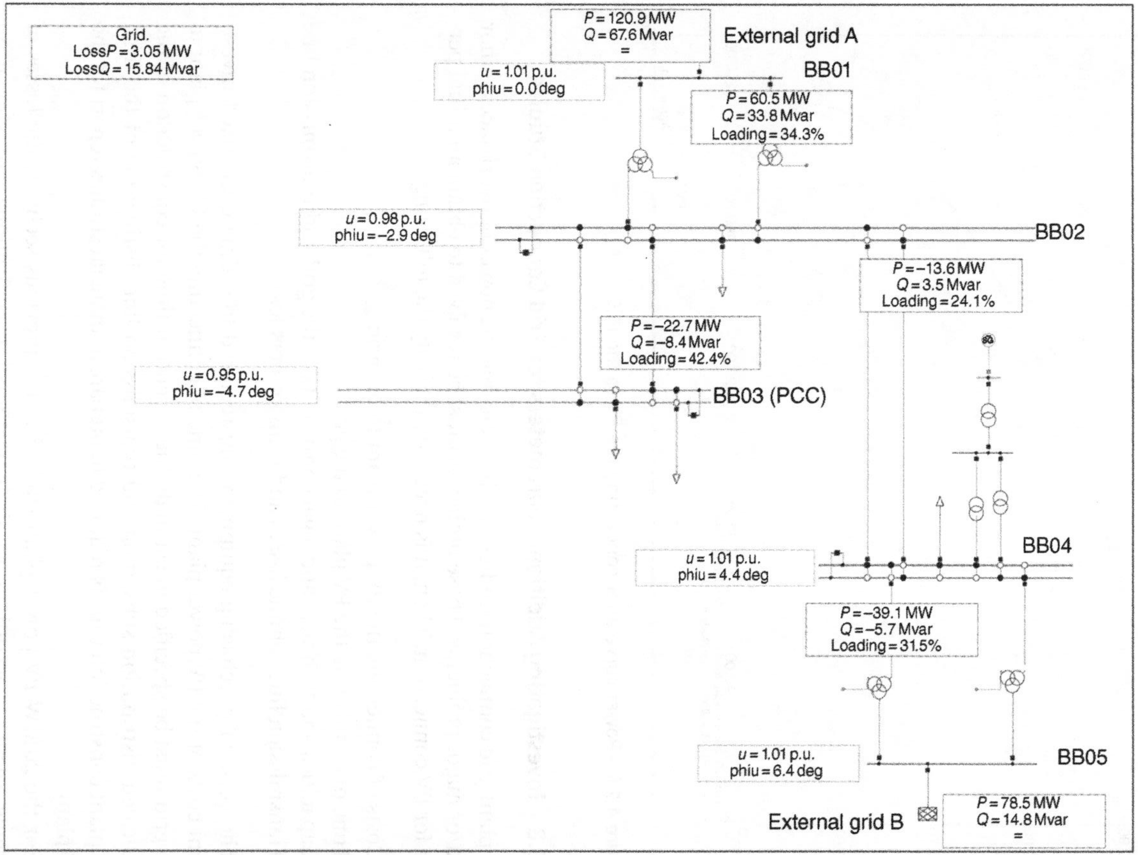
Grid.
LossP = 3.05 MW
LossQ = 15.84 Mvar
P = 120.9 MW
Q = 67.6 Mvar
=
External grid A
u = 1.01 p.u.
phiu = 0.0 deg
BB01
P = 60.5 MW
Q = 33.8 Mvar
Loading = 34.3%
u = 0.98 p.u.
phiu = –2.9 deg
BB02
P = –13.6 MW
Q = 3.5 Mvar
Loading = 24.1%
P = –22.7 MW
Q = –8.4 Mvar
Loading = 42.4%
u = 0.95 p.u.
phiu = –4.7 deg
BB03 (PCC)
BB04
u = 1.01 p.u.
phiu = 4.4 deg
P = –39.1 MW
Q = –5.7 Mvar
Loading = 31.5%
u = 1.01 p.u.
phiu = 6.4 deg
BB05
External grid B
P = 78.5 MW
Q = 14.8 Mvar
=

Figure 4.14 The network losses before connecting the PV to the PCC.

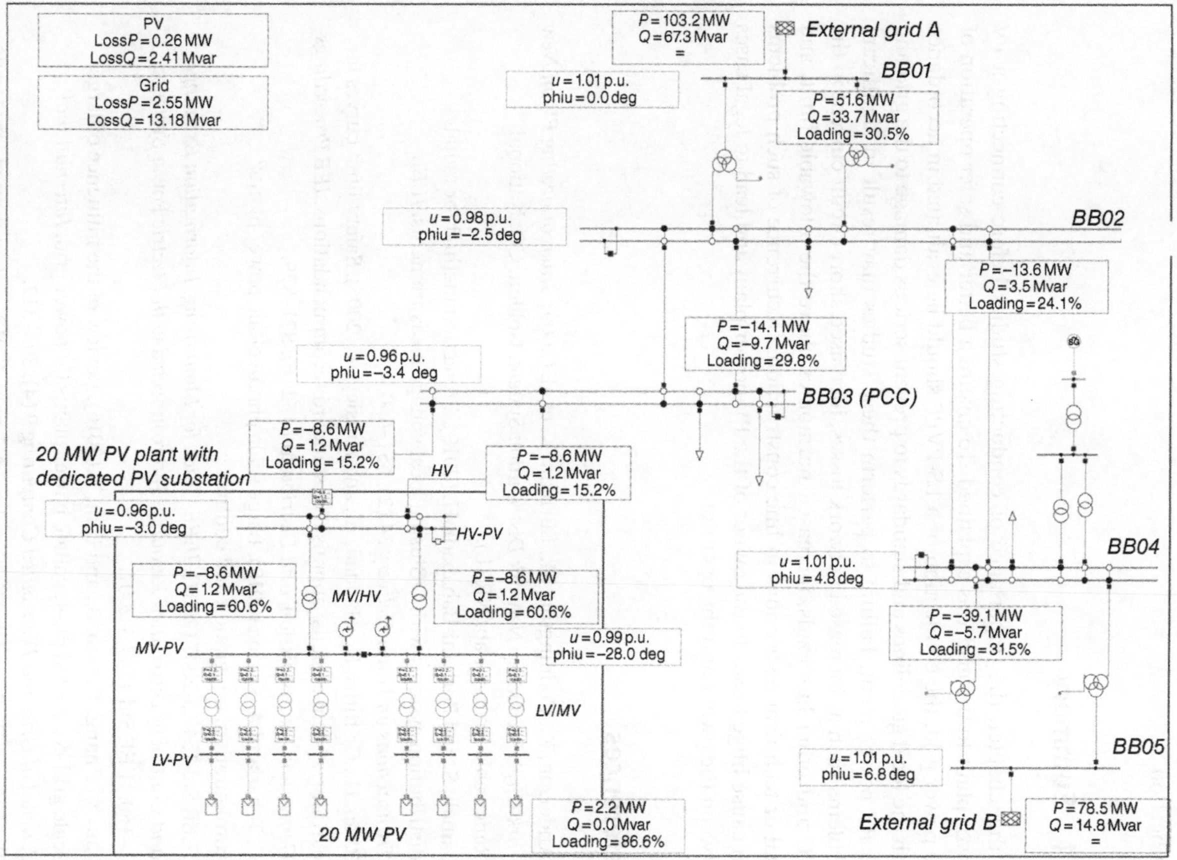
PV
LossP = 0.26 MW
LossQ = 2.41 Mvar
Grid
LossP = 2.55 MW
LossQ = 13.18 Mvar
P = 103.2 MW
Q = 67.3 Mvar
=
External grid A
BB01
u = 1.01 p.u.
phiu = 0.0 deg
P = 51.6 MW
Q = 33.7 Mvar
Loading = 30.5%
u = 0.98 p.u.
phiu = −2.5 deg
BB02
P = −13.6 MW
Q = 3.5 Mvar
Loading = 24.1%
P = −14.1 MW
Q = −9.7 Mvar
Loading = 29.8%
u = 0.96 p.u.
phiu = −3.4 deg
BB03 (PCC)
P = −8.6 MW
Q = 1.2 Mvar
Loading = 15.2%
20 MW PV plant with
dedicated PV substation
HV
P = −8.6 MW
Q = 1.2 Mvar
Loading = 15.2%
u = 0.96 p.u.
phiu = −3.0 deg
HV-PV
P = −8.6 MW
Q = 1.2 Mvar
Loading = 60.6%
MV/HV
P = −8.6 MW
Q = 1.2 Mvar
Loading = 60.6%
u = 0.99 p.u.
phiu = −28.0 deg
MV-PV
LV/MV
LV-PV
20 MW PV
P = 2.2 MW
Q = 0.0 Mvar
Loading = 86.6%
u = 1.01 p.u.
phiu = 4.8 deg
BB04
P = −39.1 MW
Q = −5.7 Mvar
Loading = 31.5%
u = 1.01 p.u.
phiu = 6.8 deg
BB05
External grid B
P = 78.5 MW
Q = 14.8 Mvar
=

Figure 4.15 The network losses after connecting the PV to the PCC.

connecting the PV power plant to the grid is 3.05 MW, while it is 2.55 MW after the PV connection, which indicates that the network losses are reduced after PV connection.

4.4 Summary

In this chapter, the importance of conducting studies before connecting a PV power plant to the grid was explained. To ensure a better and safer operation of the power grid, the connection of a LS-PVPP should be evaluated in accordance with the local guidelines and standards to prevent serious damage to the grid and the PV power plant. Failure to perform these studies may result in significant problems such as increased network losses, increased short circuit current in the PCC and other bus loads, increased harmonics above the allowable limit, and legal or technical restrictions in line construction. Occurrence of such problems can cause huge losses to the owner of the PV power plant and lead to legal cases between the owner and the operator.

References

1 Anderson, K., Coddington, M., Burman, K. et al. (2009). *Interconnecting PV on New York City's Secondary Network Distribution System*. Golden, CO: National Renewable Energy Lab.(NREL).
2 Jamali, S. and Borhani-Bahabadi, H. (2019). Protection method for radial distribution systems with DG using local voltage measurements. *IEEE Transactions on Power Delivery* 34 (2): 651–660.
3 Zhao, H., Griffiths, H., Haddad, A., and Ainsley, A. (2005). Safety-limit curves for earthing system designs: appraisal of standard recommendations. *IEE Proceedings-Generation, Transmission, and Distribution* 152 (6): 871–879.
4 C. Tobar and A. Karina (2018). Large scale photovoltaic power plants: configuration, integration and control.
5 IEEE Std 1547.3-2007 (2007). *IEEE Guide for Monitoring, Information Exchange, and Control of Distributed Resources Interconnected with Electric Power Systems*, 1–160. IEEE Std 1547.3-2007.
6 Liu, X., Zhang, T., Gao, B., and Han, Y. (2016). Review of the influence of large-scale grid-connected photovoltaic power plants on power grid. *International Journal of Grid and Distributed Computing* 9 (4): 303–312.
7 Rakhshani, E., Rouzbehi, K., Sánchez, A.J. et al. (2019). Integration of large scale PV-based generation into power systems: a survey. *Energies* 12 (8): 1425.

8 Eltawil, M.A. and Zhao, Z. (2010). Grid-connected photovoltaic power systems: technical and potential problems—a review. *Renewable and Sustainable Energy Reviews* 14 (1): 112–129.

9 Muda, H. and Jena, P. Sequence currents based adaptive protection approach for DNs with distributed energy resources. *IET Generation, Transmission & Distribution* 11 (1): 154–165. http://digital-library.theiet.org/content/journals/10.1049/iet-gtd.2016.0727.

10 Bollen, M.H. (2011). *Integration of Distributed Generation in the Power System*. John Wiley & Sons.

11 Blooming, T.M. and Carnovale, D.J. (2006). Application of IEEE Std 519-1992 harmonic limits. In: *Conference Record of 2006 Annual Pulp and Paper Industry Technical Conference*. IEEE.

5

Solar Resource and Irradiance

5.1 Introduction

The amount of solar radiation on the surface of PV modules greatly affects the produced electrical power. Sunlight is the most important meteorological factor in determining the efficiency of a PV plant. To design solar energy systems including PV plants, solar radiation data is required. Moreover, various components of solar radiation and the way they are extracted should be specified [1].

In this chapter, we discuss radiometric terms, solar resources, and solar energy radiation and its parameters including solar azimuth and altitude angle, tilt angle, shadow distances, and row spacing.

5.2 Radiometric Terms

Table 5.1 summarizes the definitions of quantities including radiation, radiation flux, radiation intensity, radiant emittance, radiance, irradiance, spectral irradiance, and irradiation [2].

5.2.1 Extraterrestrial Irradiance

Every object emits radiation at temperatures above zero. The sun behaves like a black body at a surface temperature of 5800 K. It emits extraterrestrial spectrum (ETS) radiation with wide wavelengths as shown by the ETS in Figure 5.1. The total radiant power of the sun, which is referred to as the Solar Constant (SC), is approximately constant and equal to the integral of ETS over all wavelengths. Interactions in the sun occur about every 11 years. To account for SC variability,

Step-by-Step Design of Large-Scale Photovoltaic Power Plants, First Edition. Davood Naghaviha, Hassan Nikkhajoei, and Houshang Karimi.

Table 5.1 Radiometric terminology and units [2].

Quantity	Description
Radiant energy	Energy (J)
Radiant flux	Radiant energy per unit of time/radiant power (W)
Radiant intensity	Power per unit solid angle (W/sr)
Radiant emittance	Power emitted from a surface (W/m^2)
Radiance	Power per unit solid angle per unit of projected source area (W/[m^2·sr])
Irradiance	Power incident on a unit area surface (W/m^2)
Spectral irradiance	Power incident on a unit area surface per unit wavelength measured in nanometers (W/[m^2·nm])
Irradiation	Energy accumulated on a unit area surface over a period; a more practical energy (J/m^2)

Source: Sengupta et al. [2].

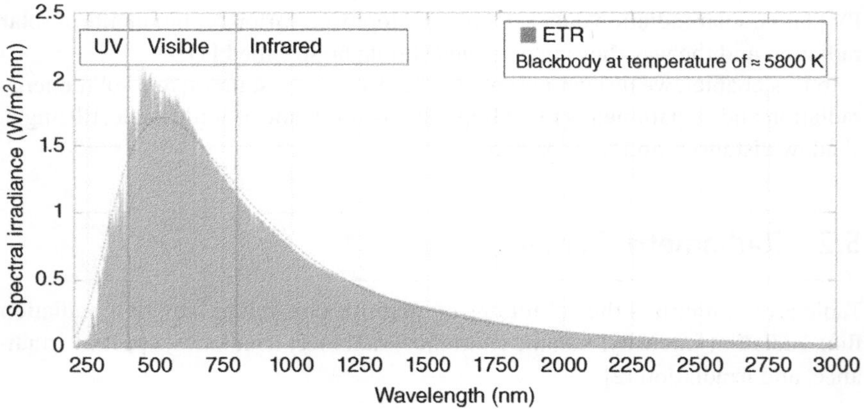
2.5
2
1.5
1
0.5
0
Spectral irradiance (W/m²/nm)
UV
Visible
Infrared
ETR
Blackbody at temperature of ≈ 5800 K
250
500
750
1000
1250
1500
1750
2000
2250
2500
2750
3000
Wavelength (nm)

Figure 5.1 Reference extraterrestrial spectrum (ASTM E-490-06) and 5800 K blackbody distribution using Planck's law. Image by Philippe Blanc, MINES-ParisTech / ARMINES. *Source:* Sengupta et al. [2].

the total radiant power at any given moment is defined and referred to as the Total Solar Irradiance (TSI) [2].

5.2.2 Solar Geometry

The distance between two objects affects the amount of radiation exchanged between them. Earth's orbit is closest to the earth in January and farthest in July. This annual change leads to *a* $\pm$3.4% change in the amount of solar radiation received by the earth.

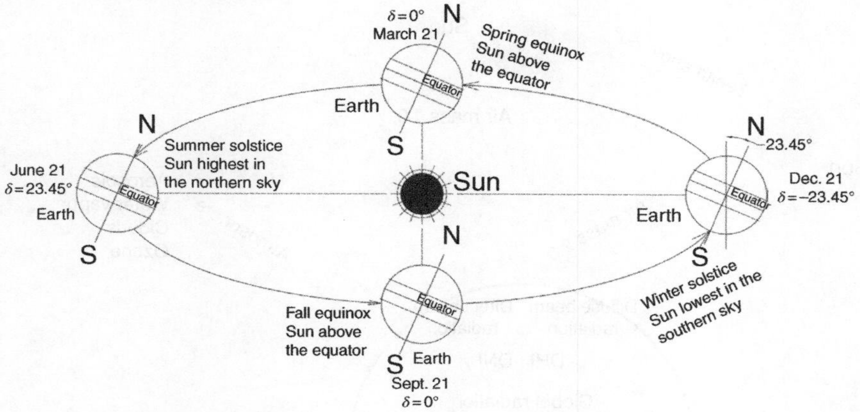
δ=0°
March 21
N
Spring equinox
Sun above
the equator
Equator
Earth
S
N
Summer solstice
Sun highest in
the northern sky
June 21
δ=23.45°
Equator
Earth
S
Sun
N
N
23.45°
Dec. 21
δ=−23.45°
Equator
Earth
S
Winter solstice
Sun lowest in the
southern sky
Equator
Fall equinox
Sun above
the equator
S
Earth
Sept. 21
δ=0°

Figure 5.2 The earth's orbit around the sun and the position of the earth's axis over a year. *Source:* Sengupta et al. [2].

Figure 5.2 shows the earth's orbit for the northern hemisphere for different seasons. The average angle of rotating axis of the earth with the orbital plane is $\delta = 23.45°$. Extraterrestrial radiation (ETR) is equal to the power per unit area emitted by the sun above the atmosphere. The ETR varies according to (5.1), where r_0 is the sun–earth distance, and r is the average annual distance [2].

$$ETR = TSI\left(\frac{r_0}{r}\right)^2 \tag{5.1}$$

5.2.3 Solar Radiation and Earth's Atmosphere

The earth's atmosphere is a variable filter for the sun's ETR. Figure 5.3 shows the absorption of solar radiation by ozone, oxygen, water vapor, and carbon dioxide. The length of atmosphere that the sun's photons travel to reach the earth's surface is called Air Mass (AM). The AM depends on the position of the earth relative to the position of the sun in the sky. The air mass is obtained from (5.2), where solar zenith angle θ is the sun angle with respect to its peak location. When the sun is at the top of the sky, the air mass is called AM1. When θ is equal to 60° (Figure 5.3), the air mass is called AM2. The atmospheric trajectory is twice the length of AM1 [2].

$$AM = \frac{1}{\cos\theta} \tag{5.2}$$

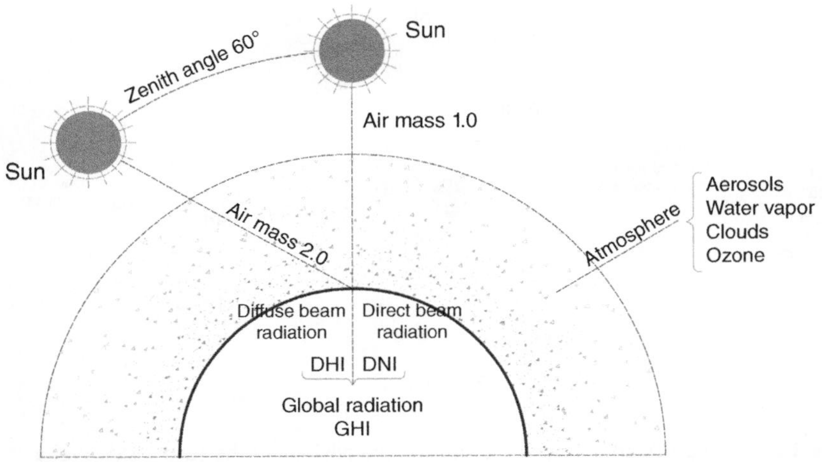
Sun
Zenith angle 60°
Sun
Air mass 1.0
Air mass 2.0
Atmosphere
Aerosols
Water vapor
Clouds
Ozone
Diffuse beam radiation
Direct beam radiation
DHI
DNI
Global radiation
GHI

Figure 5.3 Scattering of the direct-beam photons by the atmosphere: variations of AM with respect to zenith angle. Image by NREL. *Source:* Sengupta et al. [2]. Public domain.

5.3 Solar Resources

The study of solar resources provides valuable information about the amount of radiation hit on a PV plant site throughout a year. The higher the amount of annual radiation hit on the site, the higher the energy efficiency per kilowatt hour of the PV power plant and the greater the plant revenue. To predict the performance of a PV power plant, a historical data of the solar source is required. Since daily and annual radiations are variable parameters, it is preferred to use long-term (e.g. 5–10 years) radiation data to ensure an optimum plant design. However, there are PV plants that have been designed based on short-term (one to two years) radiation data.

The sunlight can be transmitted, absorbed, or diffused as it passes through the atmosphere, as shown in Figure 5.4. The solar radiation on the earth's surface consists of two components: Direct Normal Irradiance and Diffuse Horizontal Irradiance. The solar source for a PV plant site is usually defined by the parameters such as Global Horizontal Irradiation (GHI), Direct Normal Irradiation (DNI), and Diffuse Horizontal Irradiation (DHI). The geometric sum of DNI and DHI, as expressed in (5.3), is called GHI (see Figure 5.5) [4].

$$GHI = DHI + DNI\cos(\theta) \tag{5.3}$$

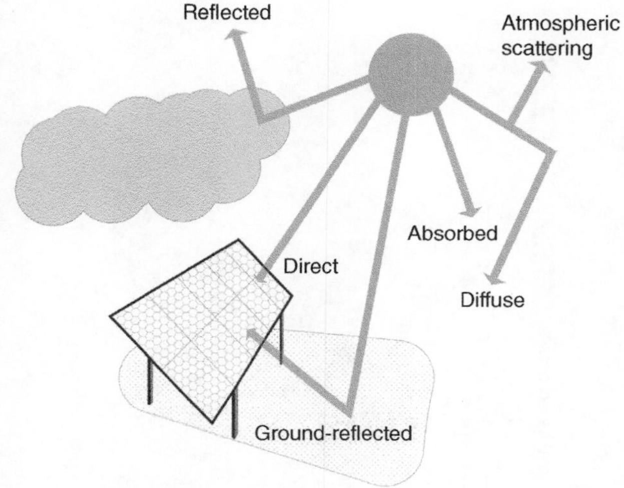
Reflected
Atmospheric scattering
Absorbed
Direct
Diffuse
Ground-reflected

Figure 5.4 Solar radiation components resulting from interactions with the atmosphere.

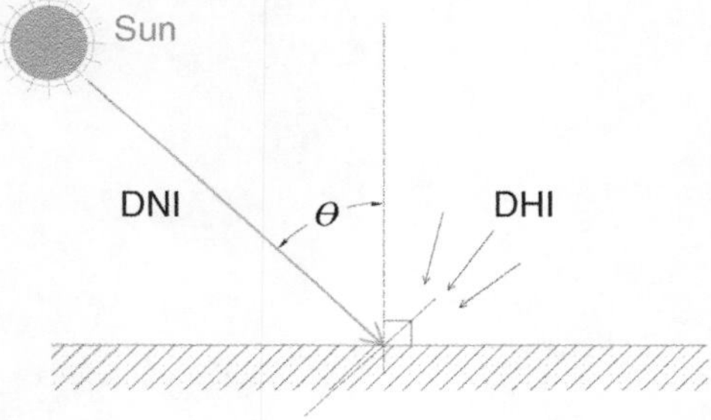
Sun
DNI
θ
DHI

Figure 5.5 Geometric representation of DHI and DNI.

- DNI is defined as the radiation per unit area. It is used for the concentrated PV, concentrated solar power, and the fixed PV installations.
- DHI is defined as the radiation emitted by the air atmosphere on a horizontal surface unit. It is used for the fixed PV installations and redundancy calculations of GHI.
- GHI is defined as the total radiation per unit horizontal area. It is used not only for the fixed PV installations but also employed to measure, correlate, and predict evaluations based on comparisons with the solar database.
- Global Tilted Irradiation (GTI) is the total radiation received by a tilted surface.

The components of solar radiation are presented in solar energy Atlases. The global distributions of GHI and DNI are shown in Figures 5.6 and 5.7. As it can be seen from the figures, the PV power generation is strongly correlated with the solar radiation. The places close to the equator have higher solar irradiations. Of course,

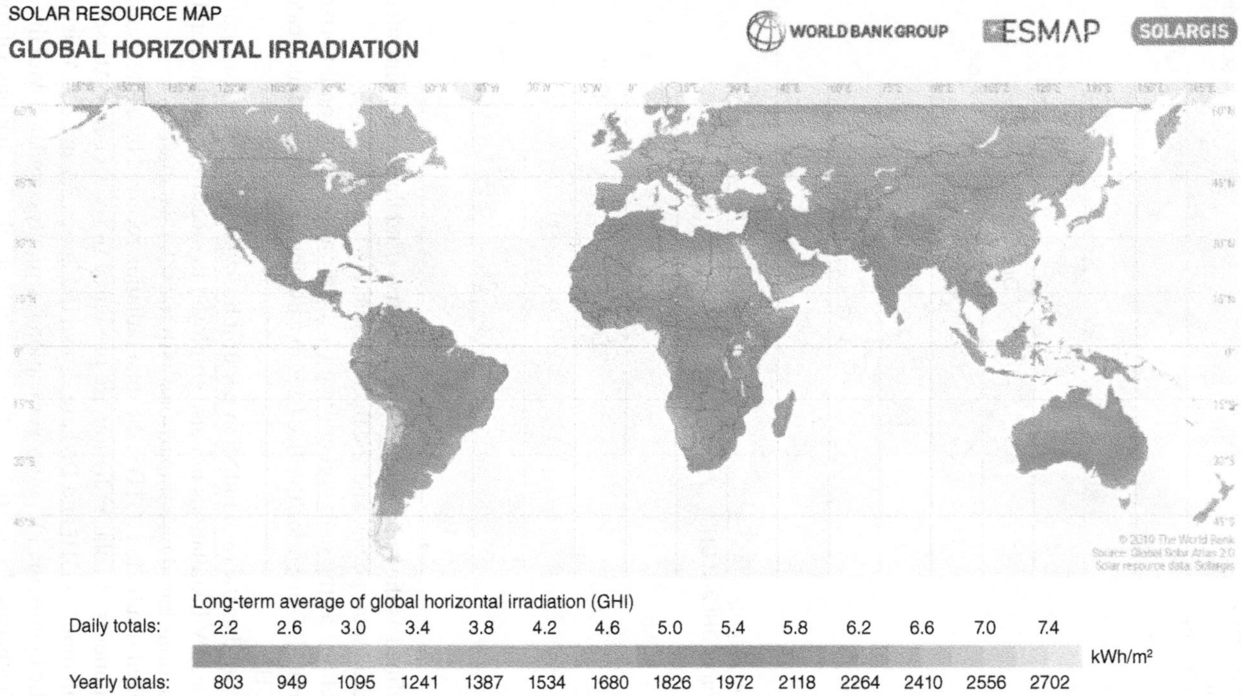
SOLAR RESOURCE MAP
GLOBAL HORIZONTAL IRRADIATION
WORLD BANK GROUP
ESMAP
SOLARGIS
Long-term average of global horizontal irradiation (GHI)
Daily totals: 2.2 2.6 3.0 3.4 3.8 4.2 4.6 5.0 5.4 5.8 6.2 6.6 7.0 7.4
kWh/m²
Yearly totals: 803 949 1095 1241 1387 1534 1680 1826 1972 2118 2264 2410 2556 2702

Figure 5.6 The worldwide spatial distributions of GHI [5].

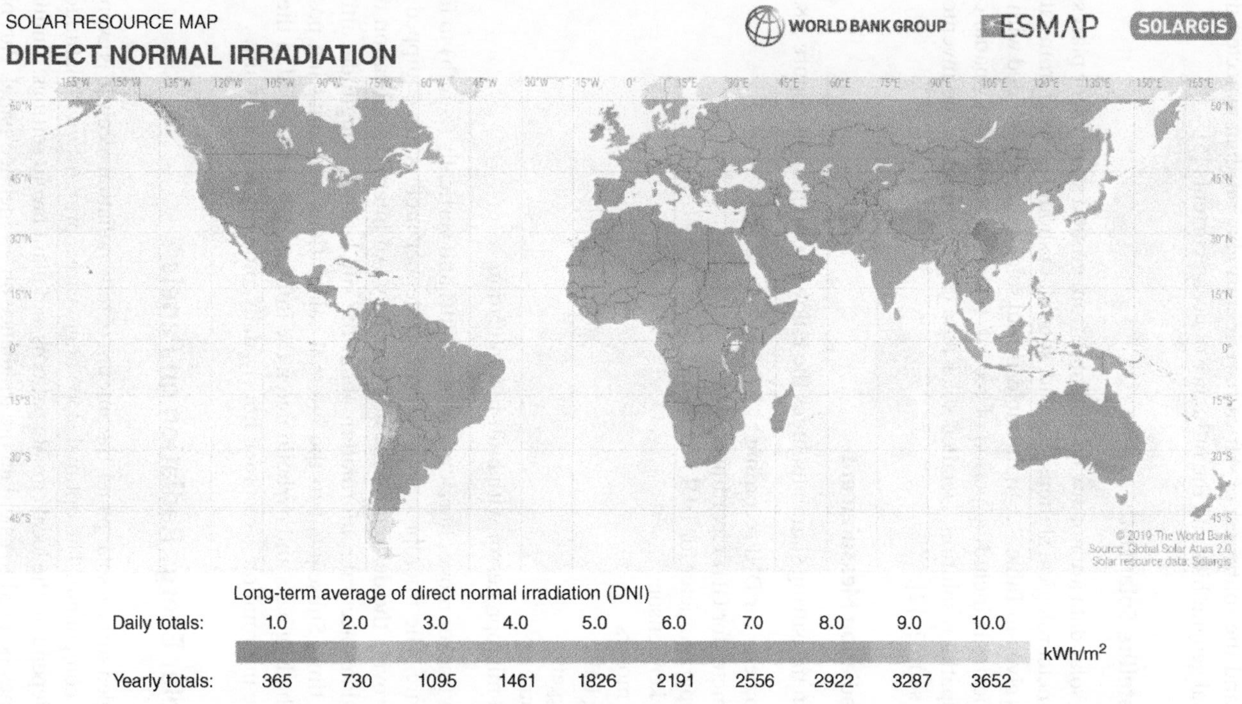
SOLAR RESOURCE MAP
DIRECT NORMAL IRRADIATION
WORLD BANK GROUP
ESMAP
SOLARGIS
© 2019 The World Bank
Source: Global Solar Atlas 2.0
Solar resource data: Solargis
Long-term average of direct normal irradiation (DNI)
Daily totals: 1.0 2.0 3.0 4.0 5.0 6.0 7.0 8.0 9.0 10.0
Yearly totals: 365 730 1095 1461 1826 2191 2556 2922 3287 3652
kWh/m²

Figure 5.7 The worldwide spatial distributions of DNI captured by [5].

there are some exceptions. For instance, the north of Australia, the west of South America, and the southwest of Africa have less irradiation than expected due to their special geographies and the influence of ocean currents [5].

5.3.1 Satellite Solar Data

Historical solar data are not always available from ground stations for places with scattered meteorological stations. Instead, solar data should be used from satellite stations as given in Table 5.2. In the table, useful solar data are provided with their key specifications including period of record, temporal resolution, spatial coverage, and spatial resolution. Detailed solar data can be found in the inventory of solar data sources [2].

5.3.2 Radiation Measurement

A radiation measuring station includes the following equipment, Figure 5.8.

- Pyrheliometer for DNI recording
- Pyranometer for GHI recording
- Shaded pyranometer for DHI recording
- Sun tracker/sensor
- Stepper motors
- Shading balls
- Data logger
- Thermometer
- Internal microprocessor with built-in application

The sun tracker moves horizontally (azimuth) and vertically (zenith) to follow the solar arc. The microprocessor-controlled stepper motors are equipped with gears to provide the desired torque and accuracy. An internal application measures/calculates accurate information about the site latitude, longitude, altitude, date, and time. Subsequent to the internal calculations and adjusting position with the shading balls and pyrheliometer, the information is recorded in the data logger. The information is recorded hourly, daily, monthly, and annually.

5.4 Solar Energy Radiation on Panels

For an angled surface solar panel, the amount of total radiation received is expressed by the H_G component. The optimal slope angle varies primarily with latitude and may also depend on the local climate patterns and plant layout settings. Simulation software may be used to calculate the irradiance on an inclined panel. Part of this calculation will take into account the radiation reflected from the ground toward the panel.

Table 5.2 Inventory of solar resource data sources, presented in alphabetical order [2].

Database	Data elements and sources	Availability
3TIER	GHI, DNI, and DHI/satellite remote-sensing input data	http://www.3tier.com
ESRA	GHI, DNI, and DHI, sunshine duration, air temperatures, precipitation, water vapor pressure, and air pressure	http://www.esraeurope.org
METEONORM	GHI, temperature, humidity, precipitation, wind speed and direction, and bright sunshine duration. GHI, DNI, DHI, GTI.	http://www.meteonorm.com
NASA Surface Meteorology and Solar Energy	GHI, DNI, and DHI from a satellite remote-sensing model. Also available: estimates of clear-sky GHI, DNI, and DHI, etc.	http://www.power.larc.nasa.gov
NOAA Network	GHI, DNI, DHI (7 stations), air temperature, relative humidity, cloud amounts, barometric pressure, wind speed and direction at 10 m, precipitation, snow cover, and weather codes.	http://www.ncdc.noaa.gov
NREL's Solar Radiation Research Laboratory Data Center (MIDC)	GHI, DNI, DHI, global on tilted surfaces, reflected solar irradiance, ultraviolet, infrared, photometric and spectral radiometers, sky imagery, and surface meteorological conditions	http://www.midcdmz.nrel.gov
PVGIS	GHI, DNI, DHI, and GTI, based on the CM-SAF database, optional terrain shadowing	http://www.ec.europa.eu
SolarGIS	DNI, GHI, DHI, GTI, and air temperature (2-m AGL) and others	http://www.globalsolaratlas.info

Source: Sengupta et al. [2].

The total energy radiated on the surface of a panel is equal to the sum of energy received from direct radiation, diffused radiation, and reflective radiation from the ground and is expressed by (5.4) as follows, Figure 5.9 [3].

$$H_G = H_{GB} + H_{GD} + H_{GR} = k_B.R_B.\left(H - H_D\right) + R_D.H_D + R_R.\rho.H \tag{5.4}$$

where H is the irradiation from global radiation on the horizontal plane, H_D is the irradiation from diffuse radiation on the horizontal plane, k_B is shading correction factor (for Non-shaded: $k_B = 1$), R_B is the direct beam irradiation factor as given in Table 5.3. R_D is the diffuse radiation factor which is obtained from

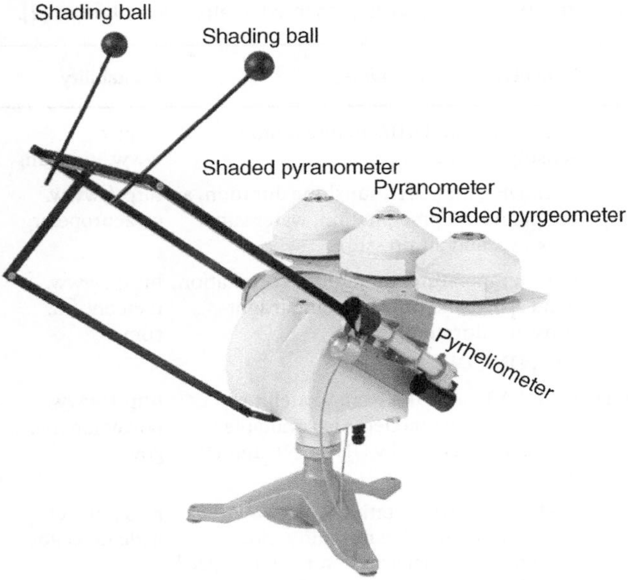
Shading ball
Shading ball
Shaded pyranometer
Pyranometer
Shaded pyrgeometer
Pyrheliometer

Figure 5.8 Automatic solar tracker. *Source:* KIPP $ ZONEN, OTT HydroMet B.V.

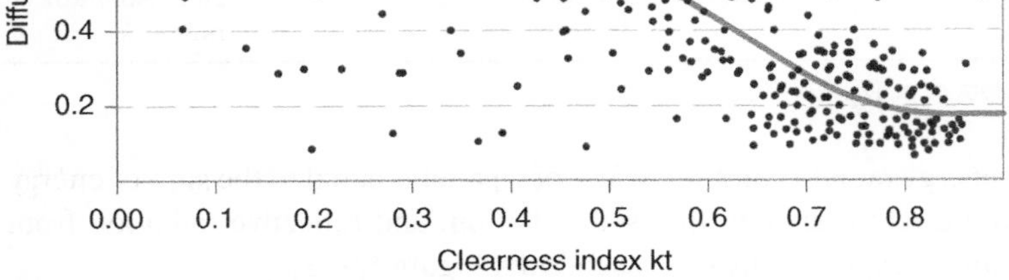
Erbs model
Measurments
1.0
0.8
0.6
0.4
0.2
Diffuse/global
0.00
0.1
0.2
0.3
0.4
0.5
0.6
0.7
0.8
Clearness index kt

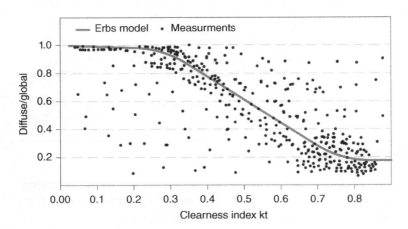

Figure 5.9 Irradiation H_G arriving at an angle of incidence relative to a horizontal plane is composed of direct beam radiation, diffuse solar radiation, and diffuse reflected radiation. *Source:* Häberlin [3].

Table 5.3 Selected mean monthly R_B values for southerly oriented surfaces [3].

Beam irradiation factors $R_B(\beta,\gamma)$ for tilted planes facing due south ($\gamma = 0°$)													
Latitude	P	Jan	Feb	Mar	April	May	June	July	Aug	Sept	Oct	Nov	Dec
$\varphi = 47°$ N	30°	2.25	1.80	1.44	1.20	1.06	1.00	1.03	1.13	1.33	1.66	2.09	2.46
	45°	2.66	2.02	1.53	1.18	0.98	0.90	0.94	1.09	1.37	1.83	2.44	2.96
	60°	2.89	2.11	1.50	1.08	0.84	0.75	0.79	0.97	1.31	1.87	2.62	3.26
	90°	2.76	1.86	1.16	0.68	0.41	0.32	0.36	0.55	0.93	1.59	2.45	3.19
$\varphi = 50°$ N	30°	2.49	1.92	1.51	1.24	1.08	1.02	1.05	1.17	1.38	1.76	2.29	2.77
	45°	3.00	2.20	1.62	1.24	1.02	0.94	0.97	1.13	1.44	1.97	2.71	3.40
	60°	3.31	2.33	1.61	1.15	0.89	0.79	0.83	1.02	1.40	2.04	2.96	3.80
	90°	3.24	2.11	1.29	0.76	0.47	0.37	0.41	0.61	1.04	1.78	2.84	3.81
$\varphi = 53°$ N	30°	2.82	2.08	1.58	1.28	1.11	1.05	1.08	1.20	1.44	1.88	2.54	3.22
	45°	3.47	2.42	1.72	1.30	1.06	0.97	1.01	1.18	1.52	2.14	3.08	4.04
	60°	3.88	2.60	1.74	1.22	0.94	0.83	0.88	1.08	1.50	2.25	3.40	4.58
	90°	3.90	2.42	1.44	0.84	0.53	0.42	0.46	0.68	1.15	2.02	3.35	4.71

Source: Häberlin [3].

Table 5.4 Guide values for the reflection factor ρ [3].

Type of surface	Reflection factor ρ (albedo)
Asphalt	0.1–0.15
Green forest	0.1–0.2
Wet ground	0.1–0.2
Dry ground	0.15–0.3
Grass-covered ground	0.2–0.3
Concrete	0.2–0.35
Desert sand	0.3–0.4
Old snow (depending on how dirty it is)	0.5–0.75
Newly fallen snow	0.75–0.9

Source: Häberlin [3].

$R_D = \frac{1}{2}\cos\alpha_2 + \frac{1}{2}\cos(\alpha_1 + \beta)$. R_R is the effective portion of reflective radiation and is calculated from $R_R = \frac{1}{2} - \frac{1}{2}\cos\beta$, where α_1 is horizon elevation in the γ direction, α_2 is facade/roof edge elevation relative to the solar generator plane, β is inclination angle of the surface relative to the horizontal plane, and ρ is reflection factor of the ground in front of the solar generator.

The reflection coefficient ρ, which is determined by the type of surface, varies from zero to one. This coefficient is often determined depending on whether the ground is dry or wet. Approximate values for the reflection coefficient ρ are presented in Table 5.4.

The direct beam irradiation factor R_B is a function of latitude φ, solar generator inclination angle β, and the solar generator azimuthγ. Mean monthly values of R_B for southerly-oriented surfaces are given in Table 5.3. Further information on the components of solar radiation is available in IEC 61724-1 and NREL/TP-5D00-68886 [2].

5.5 Solar Azimuth and Altitude Angle

The position of the sun in the sky relative to a place on the earth's surface is determined by two angles as shown in Figure 5.10. These angles are called altitude angle and azimuth angle γ. Table 5.5 presents the elements of a geometric shape shown in Figure 5.10. It also provides the other parameters required for the calculation of solar angles.

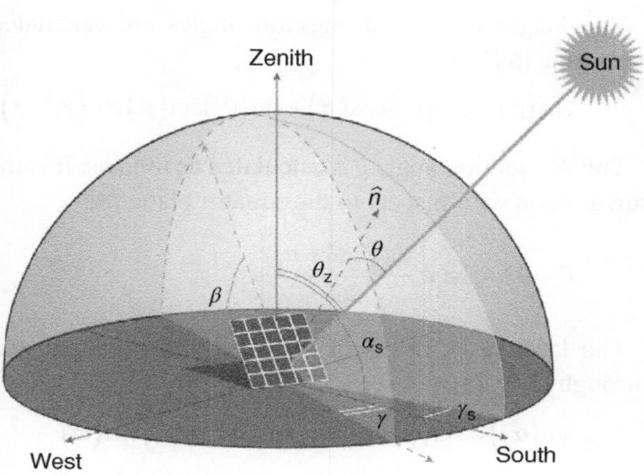
Zenith
Sun
$\hat{n}$
θ
θ_z
β
α_s
γ
γ_s
West
South

Figure 5.10 Illustration of the solar angles: (a) altitude angle, α; (b) azimuthal angle, γ. *Source:* Bhandari [6].

Table 5.5 The elements of the geometric shape shown in Figure 5.10.

φ	Latitude, the angular location north or south of the equator, north positive; $-90° \leq \varphi \leq 90°$
δ	Declination, the angular position of the sun at solar noon (i.e. when the sun is on the local meridian) with respect to the plane of the equator, north positive; $-23.45° \leq \delta \leq 23.45°$
β	Slope, the angle between the plane of the surface in question and the horizontal; $0° \leq \beta \leq 180°$ ($\beta > 90°$ means that the surface has a downward-facing component)
γ	Surface azimuth angle, the deviation of the projection on a horizontal plane of the normal to the surface from the local meridian, with zero due south, east negative, and west positive; $-180° \leq \gamma \leq 180°$
ω	Hour angle, the angular displacement of the sun east or west of the local meridian due to rotation of the earth on its axis at 15°/h; morning negative, afternoon positive
θ	Angle of incidence, the angle between the beam radiation on a surface and the normal to that surface
θ_z	Zenith angle, the angle between the vertical and the line to the sun, that is the angle of incidence of beam radiation on a horizontal surface
α_s	Solar altitude angle, the angle between the horizontal and the line to the sun, that is the complement of the zenith angle
γ_s	Solar azimuth angle: the angular displacement from south of the projection of beam radiation on the horizontal plane. Displacements east of south are negative and west of south are positive.

The height angle and azimuth angles are calculated using (5.5)–(5.7), and according to [7–9].

$$\cos(\theta) = \cos(\theta_z)\cos(\beta) + \sin(\theta_z)\sin(\beta)\cos(\gamma_s - \gamma) \tag{5.5}$$

The declination angle δ is calculated as follows. It is the angular position of the sun at noon with respect to the equator plane [8].

$$\delta = 23.45.\sin\left[\frac{2\pi(284+n)}{365}\right] \tag{5.6}$$

The latitude ϕ, the hour angle ω, and the solar zenith angle θ are related through (5.7), [8].

$$\sin(\alpha_s) = \sin(\phi)\sin(\delta) + \cos(\phi)\cos(\delta)\cos(\omega) \tag{5.7}$$

And

$$\sin(\gamma_s) = \left[\frac{\cos(\delta)\cos(\omega)}{\cos(\alpha_s)}\right] \tag{5.8}$$

Details of the algorithm for calculating the zenith angle can be found in [7]. Considering that the earth rotates 360° every 24 hours, the hour angle can be obtained as follows:

$$\omega = \frac{15^\circ}{\text{hour}}\left(\text{ST in hours before solar noon}\right) \tag{5.9}$$

The solar time is different than the clock (civil) time [8]. The solar time is expressed in terms of the clock time (CT) as follows:

$$ST - CT = 4\left(L_{ST} - L_{loc}\right) + TE \tag{5.10}$$

where L_{ST} is the standard meridian for the local time zone, L_{loc} is the location longitude, and TE is the time. L_{loc} and L_{ST} are expressed in degrees and the unit for ST and CT is minutes [8]. TE is calculated as follows.

$$TE = 9.87.\sin\left[\frac{4\pi(n-81)}{365}\right] + 7.67\sin\left[\frac{2\pi(n-1)}{365}\right] \tag{5.11}$$

where n is the day number.

5.6 Tilt Angle and Orientation

Orientation and tilt angle of PV arrays are among the most important parameters that affect the efficiency of a PV power plant. In the fixed angle PV systems, these

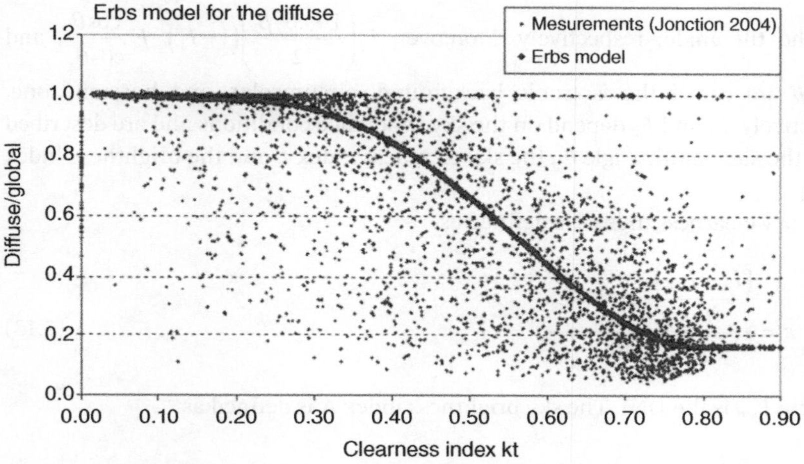

Erbs model for the diffuse
Mesurements (Jonction 2004)
Erbs model
Diffuse/global
1.2
1.0
0.8
0.6
0.4
0.2
0.0
0.00
0.10
0.20
0.30
0.40
0.50
0.60
0.70
0.80
0.90
Clearness index kt

Figure 5.11 Erbs model for the diffuse. *Source:* Based on Khoo et al. [10].

parameters must be optimally designed in order to maximize the amount of radiation received from the sun [10].

To determine the optimal installation angle of the PV module, the relationship between the PV slope angle and the amount of radiation received from the sun should be determined. Although direct radiation on a sloping surface (DNI) can be converted to a slope angle using geometric relationships, converting the diffuse radiation (DHI) to a slope angle is a more complex process. An example that represents the diffuse irradiation changes in terms of sky clearance index in terms of measured data is shown in Figure 5.11 [10].

As a rule of thumb, the angle of a panel is usually equal to the latitude of the site, and the modules are oriented along the equator. However, due to the effect of local climatic conditions on the angle and direction of the modules, this rule does not necessarily lead to an optimal design [12].

A review of the methods for optimal calculation of the slope angle is presented in [13]. Transposition models can be used to simulate the amount of radiation on PV panels for different azimuth angles and slopes using GHI and DHI measured data. The Perez model presented in [10] is one of the most accurate and commonly used one. In this model, the total tilted diffuse irradiance, $I_{d,tilt}$, is given by

$$I_{d,tilt} = I_d\left(\frac{1+\cos\beta}{2}\right)(1-F_1) + F_1\frac{\cos\theta}{\cos\theta_z} + F_2\sin\beta \tag{5.12}$$

In (5.12), θ, θ_z, I_d, F_1, F_2, *and* β are the angle of incidence, the zenith angle, the DHI, the circumsolar brightness coefficient, the horizon brightness coefficient, and the tilt angle, respectively. Moreover, $I_d\left(\frac{1+\cos\beta}{2}\right)(1-F_1)$, $F_1\frac{\cos\theta}{\cos\theta_z}$, and $F_2\sin\beta$ are called the isotropic background, circumsolar, and horizon zone, respectively. F_1 and F_2 depend on the sky irradiance conditions and are described using the sun zenith angle θ_z, the sky clearness index ε, and the brightness index Δ [10].

The sky clearness index ε is defined as

$$\varepsilon = \frac{\frac{(I_d + I_{b,n})}{I_d} + 1.041\theta_z^3}{1 + 1.041\theta_z^3} \tag{5.13}$$

where $I_{b,n}$ is the DNI. The sky brightness index Δ is defined as

$$\Delta = m\frac{I_d}{I_E} \tag{5.14}$$

where m is the air mass and I_E is the extraterrestrial irradiance. The sky clearness index ε is separated into eight bins. In each bin, the brightness coefficients F_1 and F_2 are linear functions of θ_z and ε as

$$F_1 = f_{11}(\varepsilon) + \Delta f_{12}(\varepsilon) + \theta_z f_{13}(\varepsilon) \tag{5.15}$$

$$F_2 = f_{21}(\varepsilon) + \Delta f_{22}(\varepsilon) + \theta_z f_{23}(\varepsilon) \tag{5.16}$$

where the coefficients F_{ij} are determined using the least square algorithm to fit with the experimental data. The total tilted irradiance I_T is obtained by adding the beam irradiance I_b.

$$I_T = I_b R_b + I_{d,tilt} + \mathrm{I}\rho_g\left(\frac{1-\cos\beta}{2}\right) \tag{5.17}$$

where I_b is the beam irradiance on a horizontal surface, R_b is the ratio of the beam radiation on the tilted surface to that of the horizontal surface at a given time, $I_{d,tilt}$ is the DHI, I is the global horizontal irradiance, and ρ_g is the ground reflectance [10].

The Perez model has been compared with the radiated data measured in Singapore for different angles and directions. The error percentage of the model is presented in [10]. Figure 5.12 shows the computed annual irradiation for all possible orientations based on the acceptable error of the Perez model. As observed, when the surface of the PV module is facing 97° SE with a slope angle of 26°, the panel receives the maximum annual irradiation, equivalent to 1562 kWh/m^2. A comparison of the normalized root mean square errors (NRMSE) for all different

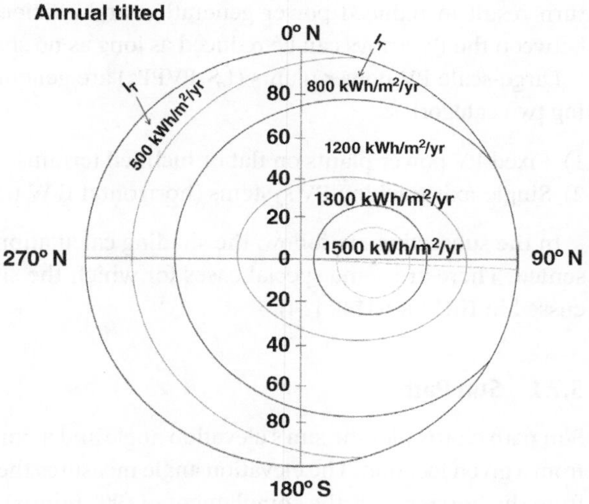
Annual tilted
0° N
800 kWh/m²/yr
500 kWh/m²/yr
1200 kWh/m²/yr
1300 kWh/m²/yr
1500 kWh/m²/yr
270° N
90° N
180° S
80
60
40
20
0
20
40
60
80

Figure 5.12 Polar contour plot of annual tilted irradiation for different tilts and orientations in Singapore.

Table 5.6 NRMSE for all different orientations and tilt angles at SERIS meteorological station in Singapore.

	Normalized root mean square errors (NRMSE)							
Azimuth	**60° NE**	**60° NE**	**60° NE**	**60° NE**	**0° N**	**180° S**	**90° E**	**270° W**
Tilt	10°	20°	30°	40°	90°	90°	90°	90°
Perez (%)	0.6	1.1	1.6	2.2	3.6	5.1	3.0	3.2

Source: Based on Duffie and William [7].

orientations and tilt angles in a meteorological station in Singapore based on the Perez Model [7] is given in Table 5.6. In the PVsyst-Version5 software, the Hay model is used to represent the emitted radiation, and a more accurate Perez model is implemented in Version6 of this software.

5.7 Shadow Distances and Row Spacing

As mentioned in the Site Selection (Section 3.3.1), the cost of land for a PV power plant is an important part of the total cost. Therefore, to reduce the total cost, the land area of the solar power plant should be minimized. This can be achieved if the distances between the PV arrays are reduced. However, the short distances between the PV arrays can lead to the shading of the front arrays on the rear ones, which in turn result in reduced power generation and efficiency. Therefore, the distance between the PV arrays can be reduced as long as no shading happens [9].

Large-scale PV power plants (LS-PVPPs) are generally divided into the following two categories:

1) Fixed PV power plants on flat or inclined terrains
2) Single-axis tracking PV systems (horizontal E-W tracking axis)

In the subsections to follow, the shading calculations for each PV type are presented. There are some special cases for which the shading calculations are discussed in IEC TR 63149 [14].

5.7.1 Sun Path

Sun path charts plot the sun's elevation angle and azimuth angle over a day as seen from a given location. The elevation angle measures the height of the sun in the sky from the horizon. It is the complement of (90° minus) the zenith angle of the sun. The azimuth angle indicates the direction of the sun in the horizontal plain from a given location. North is defined to have an azimuth of 0° and south has an azimuth of 180°. In addition to depicting the path of the sun in terms of its elevation and azimuth angles, sun path charts indicate particular times of the day [15]. Sun path diagram for any specified location on earth at any specified time of the year has been obtained by entering the relevant information at the University of Oregon's solar radiation monitoring laboratory [15].

Figure 5.13 Shows the solar path curve for the city of Farmington in New Mexico. According to this figure, on 21 December and 21 June, the height of the sun is at its lowest and highest position from the earth, respectively. On 21 December, the height of the sun from the earth in the specified region is at its shortest, indicating that the length of the shadow is longer than those of the other days of the year. Therefore, the shading calculations will be performed for the worst case scenario, being 21 December, for the geographical location under study.

Figure 5.14 shows a ground-mounted fixed PV power plant and the shading caused by the PV arrays.

5.7.2 Shadow Calculations for Fixed PV Systems

The fixed angle PV system is installed in a flat ground as shown in Figure 5.14. First, it is assumed that the solar arrays are facing south. To determine the distance between the arrays, according to Figure 5.15 and using (5.18)–(5.20), the computational model is extracted considering the boundary conditions [9].

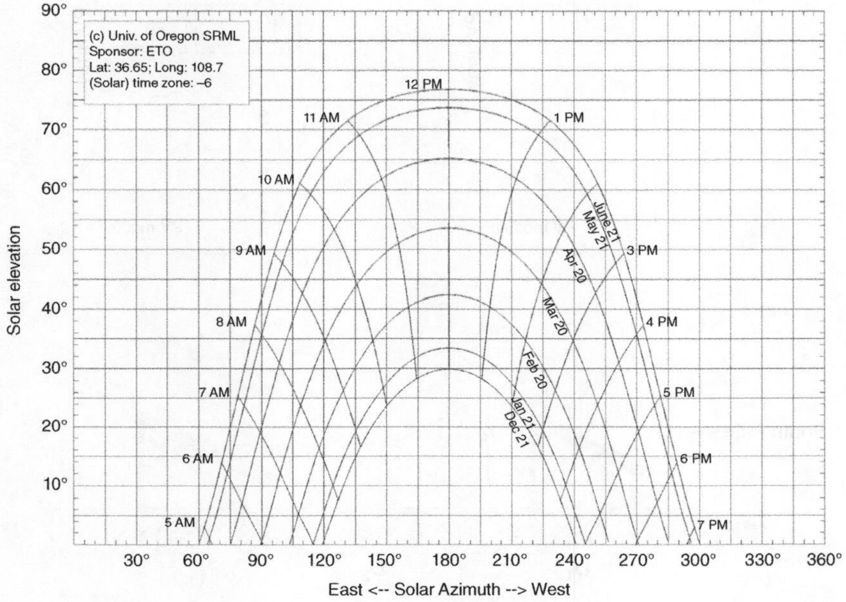
(c) Univ. of Oregon SRML
Sponsor: ETO
Lat: 36.65; Long: 108.7
(Solar) time zone: –6
Solar elevation
East <-- Solar Azimuth --> West
12 PM
11 AM
1 PM
10 AM
9 AM
3 PM
8 AM
4 PM
7 AM
5 PM
6 AM
6 PM
5 AM
7 PM
June 21
May 21
Apr 20
Mar 20
Feb 20
Jan 21
Dec 21

Figure 5.13 Sun path charts for Farmington.

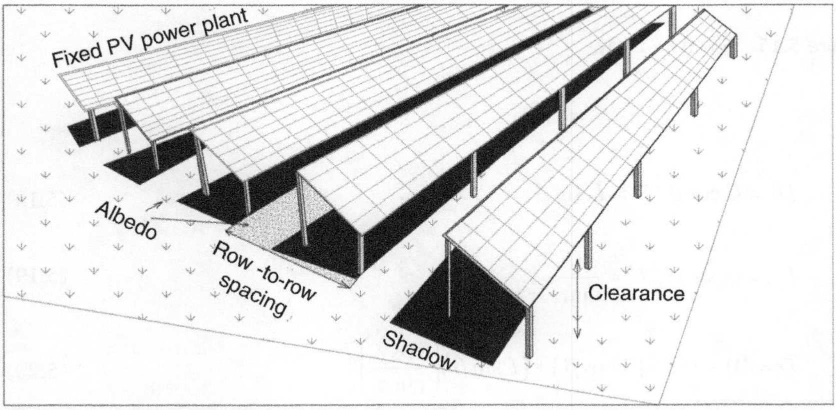
Fixed PV power plant
Albedo
Row -to-row spacing
Clearance
Shadow

Figure 5.14 Fixed ground-mounted PV array and its shading.

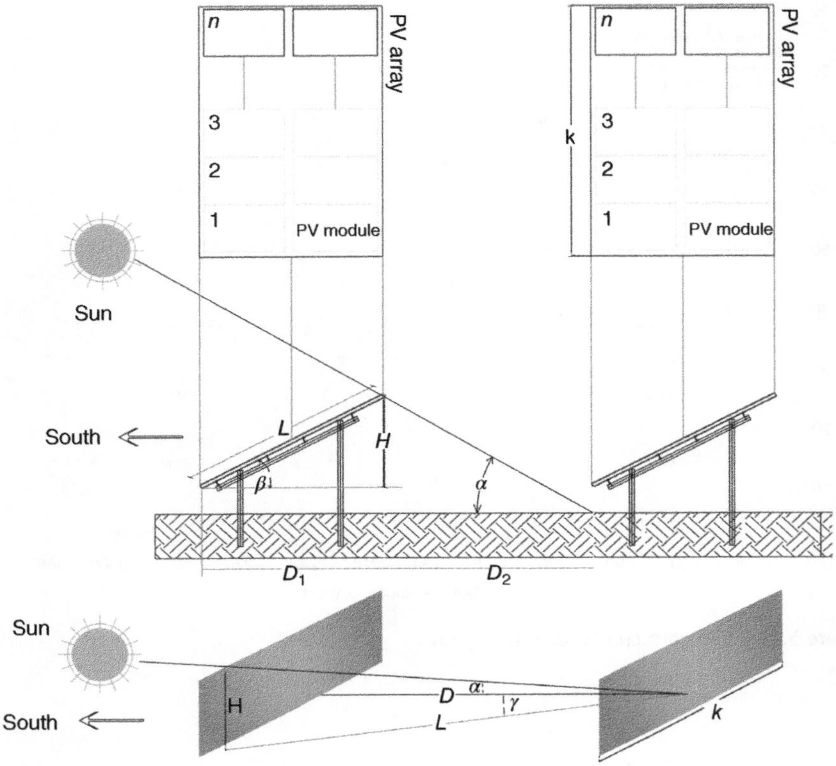
n
3
2
1
PV module
PV array
n
3
2
1
k
PV module
PV array
Sun
South
L
H
β
α
D1
D2
Sun
South
H
D
α
γ
L
k

Figure 5.15 Fixed PV array.

$$D1 = L.\cos\beta,\ H = L.\sin\beta \tag{5.18}$$

$$D2 = \cos\gamma.L',\ L' = \frac{H}{\tan\alpha} \tag{5.19}$$

$$D = D1 + D2 = \left(L\cos\beta\right) + \left(L\sin\beta\right)\left(\frac{\cos\gamma}{\tan\alpha}\right) \tag{5.20}$$

where γ is solar azimuth, α is solar altitude, D is shading distance, H is height, L is length of PV array, L' is projective length of solar rays, and β is tilt angle.

5.7.3 Shadow Calculations for Single-Axis Tracking PV Systems (Horizontal E–W Tracking Axis)

Figure 5.16 shows a single-axis tracking PV system (horizontal E–W tracking axis). Shading and spacing between the arrays for single-axis tracking PV systems are calculated from (5.21)–(5.25):

$$D1 = K.\cos A \tag{5.21}$$

$$D2 = \frac{H}{\tan\alpha} \tag{5.22}$$

$$H = K.\sin A \tag{5.23}$$

$$D = D1 + D2 = \left(K.\cos A\right) + \frac{\left(K.\sin A\right)}{\tan\alpha} \tag{5.24}$$

$$\alpha = \arcsin\left(\sin\phi\sin\delta + \cos\phi\cos\delta\cos\omega\right) \tag{5.25}$$

where A is E–W tilted angle of the PV array, α is solar altitude, D is shading distance, H is height, K is weight of PV array, φ is latitude of the location, δ is solar declination (−23.45° on winter solstice), and ω is solar hour angle. Additional information on the shading calculations and spacing between the arrays for sloping ground for the fixed-angle PV plants, and the single-axis and dual-axis PV plants is available in the IEC TR 63149 [14].

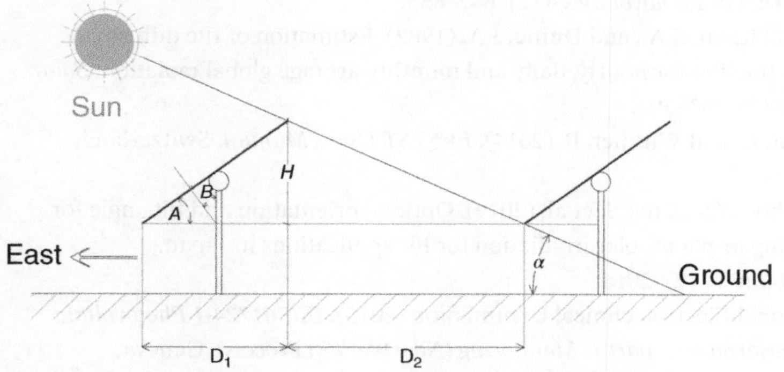
Sun
H
B
A
East
α
Ground
D_1
D_2

Figure 5.16 Single-axis tracking PV system.

References

1 Reddy, K.S. and Ranjan, M. (2003). Solar resource estimation using artificial neural networks and comparison with other correlation models. Energy Conversion and Man*agement* 44 (15): 2519–2530.

2 Sengupta, M., Habte, A., Gueymard, C., Wilbert, S., Renné, D. and Stoffel, T. (2017). Best Practices Handbook for the Collection and Use of Solar Resource Data for Solar Energy Applications: Technical Report NREL/TP.

3 Häberlin, H. (2012). *Photovoltaics: System Design and Practice*. Wiley.

4 Kumar, D.S., Yagli, G.M., Kashyap, M., and Srinivasan, D. (2020). Solar irradiance resource and forecasting: a comprehensive review. *IET Renewable Power Generation* 14 (10): 1641–1656.

5 Zhao, B., Zhao, Z., Huang, M., Zhang, X., Li, Y., and Wang, R. (2021). Model Predictive Control of Solar PV-Powered Ice-Storage Air-Conditioning System Considering Forecast Uncertainties, in *IEEE Transactions on Sustainable Energy* 12 (3): 1672–1683. doi: 10.1109/TSTE.2021.3061776.

6 Bhandari, K. (2019). The impact of tilt angle on photovoltaic panel output. *International Journal of Applied Nanotechnology* 5 (2): 39–45.

7 Duffie, J.A. and Beckman, W.A. (2013). *Solar Engineering of Thermal Processes*. Wiley.

8 Rosa-Clot, M. and Tina, G.M. (2017). *Submerged and Floating Photovoltaic Systems: Modelling, Design and Case Studies*. Academic Press.

9 IEC TR 63149 (2018). Land usage of photovoltaic (PV) farms – mathematical models and calculation examples.

10 Khoo, Y.S., Nobre, A., Malhotra, R. et al. (2013). Optimal orientation and tilt angle for maximizing in-plane solar irradiation for PV applications in Singapore. *IEEE Journal of Photovoltaics* 4 (2): 647–653.

11 Erbs, D.G., Klein, S.A., and Duffie, J.A. (1982). Estimation of the diffuse radiation fraction for hourly, daily and monthly-average global radiation. *Solar Energy* 28 (4): 293–302.

12 Mermoud, A. and Wittmer, B. (2014). *PVSYST User's Manual*. Switzerland, January.

13 Yu, C., Khoo, Y.S., Chai, J. et al. (2019). Optimal orientation and tilt angle for maximizing in-plane solar irradiation for PV applications in Japan. *Sustainability* 11 (7): 2016.

14 International Electrotechnical Commission (2015). *IEC 61724-1 Photovoltaic System Performance–part 1: Monitoring (New Work in Process)*. Geneva, Switzerland: IEC Central Office.

15 University of Oregon (2008). Solar Radiation Monitoring Laboratory. http://solardat.uoregon.edu/SunChartProgram.html (accessed 22 October 2008).

6

Large-Scale PV Plant Design Overview

6.1 Introduction

Large-scale PV power plant (LS-PVPP) projects are generally carried out by engineering, procurement, and construction (EPC) methods. Design and engineering are part of the engineering phase. In this phase, many documents and plans are produced for use in the supply and construction phase. In addition to designing different parts of the power plant with the classification of engineering documents, it is necessary for the design team to be familiar with the design methodology of an LS-PVPP. Due to its importance, in this chapter, more details of engineering documents and their classification are presented.

6.2 Classification of LS-PVPP Engineering Documents

Due to the scale of construction of LS-PVPP, the volume of documents is high and the process producing documents is complicated. The classification of LS-PVPP engineering certificates, in general, is divided into four main categories as shown in Figure 6.1. A brief description of Figure 6.1 is given below:

6.2.1 Part 1: Feasibility Study

Part 1 is the most basic phase for starting the project. In this part, feasibility study of the PV plant is performed to evaluate the feasibility of constructing a LS-PVPP at various levels. In Chapters 3 and 4, more detailed explanations of the feasibility study were provided.

Step-by-Step Design of Large-Scale Photovoltaic Power Plants, First Edition. Davood Naghaviha, Hassan Nikkhajoei, and Houshang Karimi.

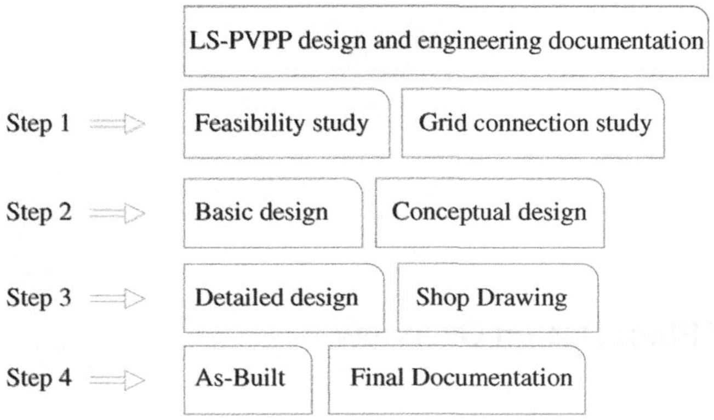
LS-PVPP design and engineering documentation
Step 1
Feasibility study
Grid connection study
Step 2
Basic design
Conceptual design
Step 3
Detailed design
Shop Drawing
Step 4
As-Built
Final Documentation

Figure 6.1 Classification of LS-PVPP engineering documents.

6.2.2 Part 2: Basic Design

In basic design, the LS-PVPP owner or stakeholders must prepare the bidding documents to select the contractor team. To prepare the documents, the initial plan of the PV plant is specified by considering various scenarios suggested by the owner in the conceptual design. The conceptual design is then presented in joint meetings between the owner, consultants, and other stakeholders. Once the best option is determined, the initial layout and plan of the PV plant is selected.

Details of the conceptual design are generally not sufficient and only provide a roadmap for the location of the main blocks and equipment of the PV plant. The evidence produced in a conceptual design is shown in Figure 6.2. Description of the sections shown in Figure 6.2 is presented as follows.

a) Master Document List

The larger the size of a PV plant project, the larger the volume of documents and drawings. In order to divide the tasks of design team and to carefully control and follow the documents, a list of engineering and design documents must first be prepared, which is called the master document list (MDL) engineering certificates. This list typically includes the items described in Figure 6.3 [1].

An MDL has an item number, a certificate number, and a certificate title. Each engineering document also has a unique certificate number. The numbering of documents in each project can be different depending on the instructions and internal standards of the employer. Document numbering for LS-PVPP projects is mainly carried out according to Figure 6.3. The titles of the documents are also shared with an engineering team who are involved in the design and engineering phase of LS-PVPP.

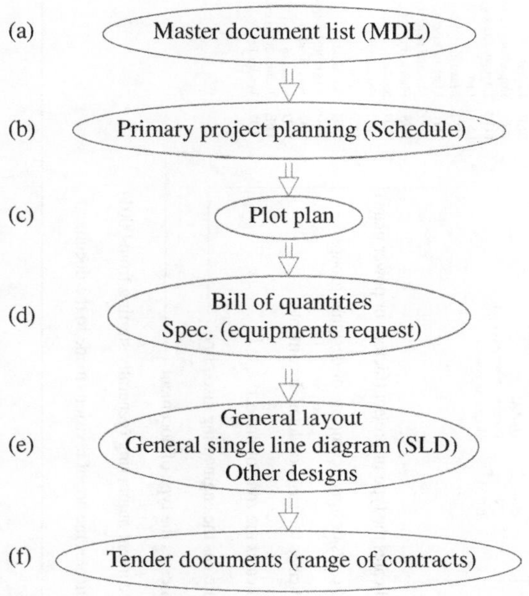
(a)
Master document list (MDL)
(b)
Primary project planning (Schedule)
(c)
Plot plan
(d)
Bill of quantities
Spec. (equipments request)
(e)
General layout
General single line diagram (SLD)
Other designs
(f)
Tender documents (range of contracts)

Figure 6.2 Documents of different parts of basic design.

Each engineering team may involve several specializations. For example, in the electrical engineering team, there are experts in protection system, ground system, cable, transformer, monitoring and other.

b) Primary Project Planning

After preparing MDL, the primary project planning must be prepared. Scheduling the certification helps manage time, cost, quality, changes, risks, and related issues. The contractor should ensure that the plant engineering and design documentation is ready on time.

Scheduling outputs are usually displayed in the form of a Gantt chart. Figure 6.4 shows an example of a Gantt chart schedule, which provides various information as below.

- Classification and list of documents or activities
- Start time of an activity
- Time scale of each activity
- Overlapping duration
- Time connection between activities
- Start and end dates of the whole project and each stage of the project
- Critical path

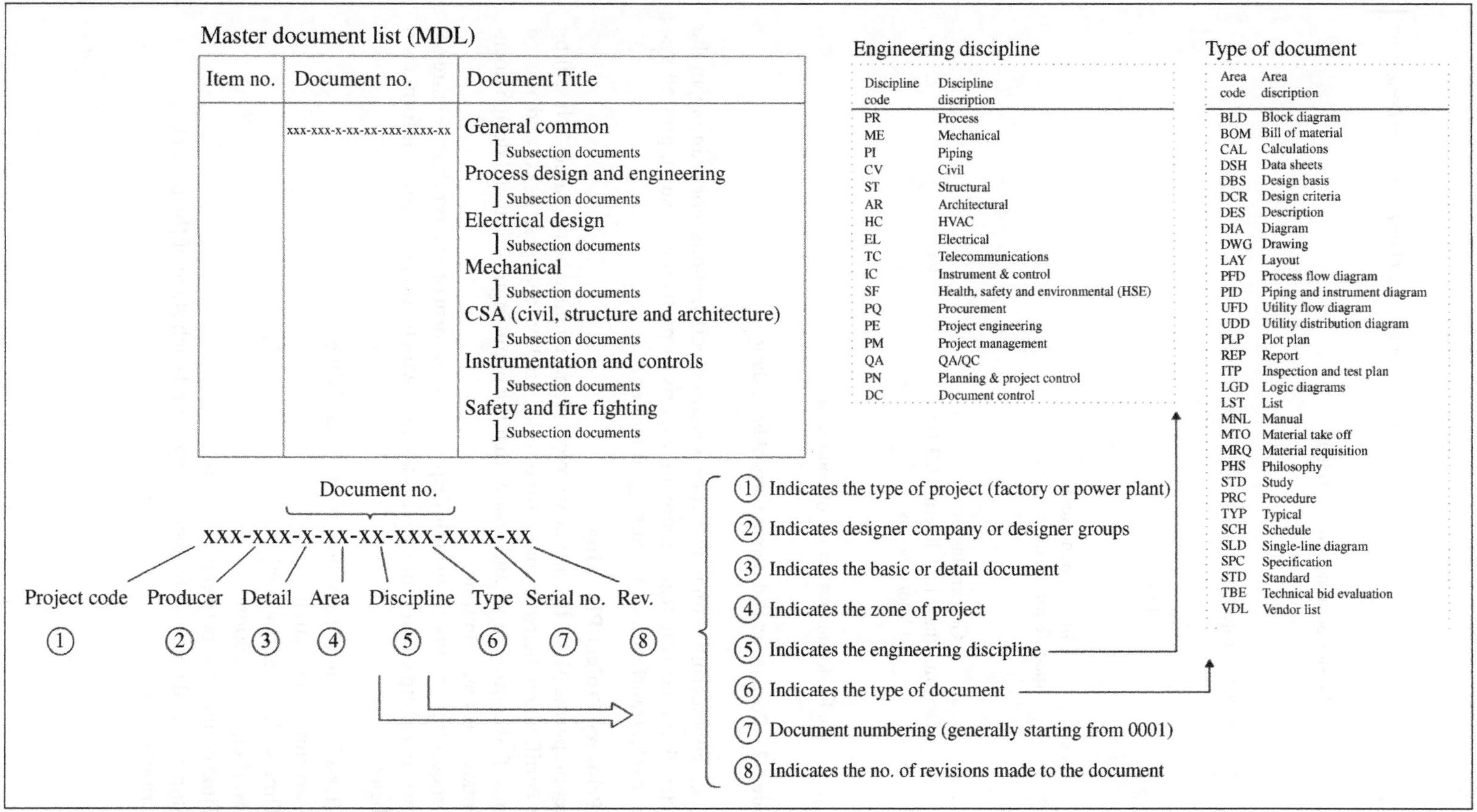
Master document list (MDL)
Item no.
Document no.
Document Title
xxx-xxx-x-xx-xx-xxx-xxxx-xx
General common
Subsection documents
Process design and engineering
Subsection documents
Electrical design
Subsection documents
Mechanical
Subsection documents
CSA (civil, structure and architecture)
Subsection documents
Instrumentation and controls
Subsection documents
Safety and fire fighting
Subsection documents
Engineering discipline
Discipline code
Discipline discription
PR Process
ME Mechanical
PI Piping
CV Civil
ST Structural
AR Architectural
HC HVAC
EL Electrical
TC Telecommunications
IC Instrument & control
SF Health, safety and environmental (HSE)
PQ Procurement
PE Project engineering
PM Project management
QA QA/QC
PN Planning & project control
DC Document control
Type of document
Area code
Area discription
BLD Block diagram
BOM Bill of material
CAL Calculations
DSH Data sheets
DBS Design basis
DCR Design criteria
DES Description
DIA Diagram
DWG Drawing
LAY Layout
PFD Process flow diagram
PID Piping and instrument diagram
UFD Utility flow diagram
UDD Utility distribution diagram
PLP Plot plan
REP Report
ITP Inspection and test plan
LGD Logic diagrams
LST List
MNL Manual
MTO Material take off
MRQ Material requisition
PHS Philosophy
STD Study
PRC Procedure
TYP Typical
SCH Schedule
SLD Single-line diagram
SPC Specification
STD Standard
TBE Technical bid evaluation
VDL Vendor list
Document no.
XXX-XXX-X-XX-XX-XXX-XXXX-XX
Project code
Producer
Detail
Area
Discipline
Type
Serial no.
Rev.
1
2
3
4
5
6
7
8
1 Indicates the type of project (factory or power plant)
2 Indicates designer company or designer groups
3 Indicates the basic or detail document
4 Indicates the zone of project
5 Indicates the engineering discipline
6 Indicates the type of document
7 Document numbering (generally starting from 0001)
8 Indicates the no. of revisions made to the document

Figure 6.3 MDL structure. *Source:* Wulff et al. [1] and Anderson et al. [2].

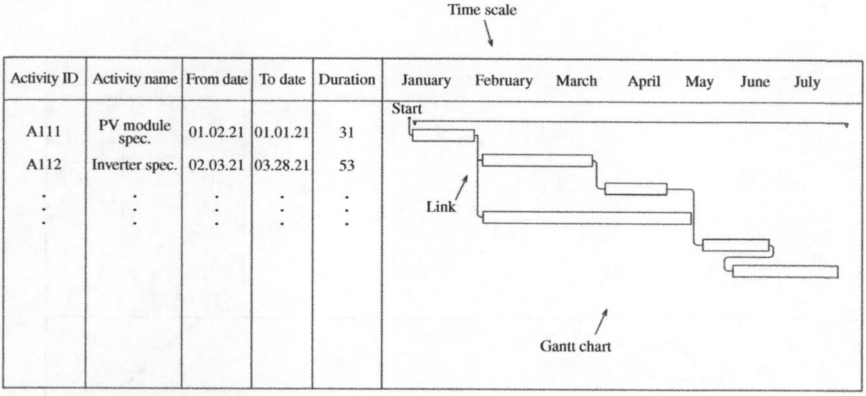
Time scale
Activity ID
Activity name
From date
To date
Duration
January
February
March
April
May
June
July
Start
A111
PV module spec.
01.02.21
01.01.21
31
A112
Inverter spec.
02.03.21
03.28.21
53
Link
Gantt chart

Figure 6.4 An example of a Gantt chart schedule.

Remarks:

- There is no prerequisite in the start of the project and no activity after the end of project.
- Except for the start and end activities of the project, all activities in the project schedule must have prerequisites and post-requirements.

c) Plot Plan

Plot plan is a map that shows general information of the site such as coordinates and boundaries of the land, land use, access roads, point of common coupling (PCC) nodes, and other useful information. This map is usually in the form of an orthomosaic map and contains details of the land. Figure 6.5 shows an example of a plot plan.

d) Bill of Quantities Definition and Specifications

In this section, small quantities and technical specifications of various PV power plant components are calculated and prepared in a list as bill of quantities (B&Q). The bill of quantities and specifications is determined based on the request of the project owner or stakeholders. For example, the number of PV modules, the number of inverters, the length of cables and other equipment along with their technical specifications are prepared for tender documents based on previous PV plants already constructed.

e) General Layout and General Single-Line Diagram

Using the basic technical specifications of the equipment, general layout and general single-line diagram (SLD) are prepared to make the owner's request more transparent to the contractors and to make the price offered more competitive between the contractors.

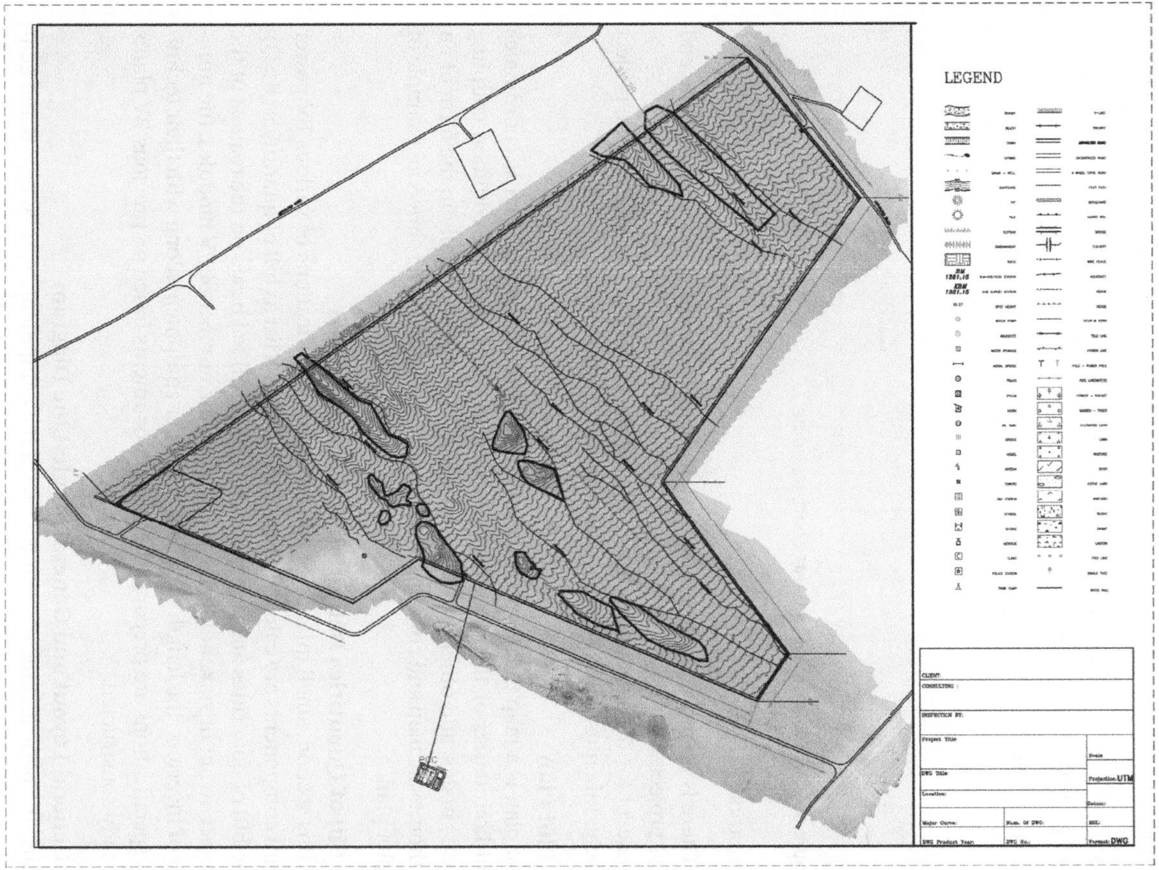
LEGEND
CLIENT:
CONSULTING :
INSPECTION BY:
Project Title:
Scale:
DWG Title:
Projection: UTM
Location:
Datum:
Major Curve:
Num. Of DWG:
MSL:
DWG Product Year:
DWG No.:
Format: DWG

Figure 6.5 An example of a plot plan.

f) Tender Documents

In this section, tender documents including forms and contracts of different parts of supply, execution, and other related to the contractor are prepared.

6.2.3 Part 3: Detailed Design and Shop Drawing

Detailed design begins when the contractor team is selected. The detailed design includes performing final calculations, preparing technical specifications of all PV plant equipment, preparing executive plans of all sections, and preparing a list of equipment materials and materials required in each part of the plant. Shop drawing are the maps that are prepared in parallel in the site.

6.2.4 Part 4: As-Built and Final Documentation

When each engineering document is produced according to procedures (predetermined), it must be sent to a supervisor or consultant for approval. Typically, an engineering document is published with one of the names shown in Figure 6.6.

Definitions of the names shown in Figure 6.6 are explained in the following sections:

a) **Issue for Information (IFI):** When an engineering certificate is only sent to a consultant or contractor for information, it is called an IFI certificate. For the IFI documents, there is no comment included.
b) **Issue for Comment (IFC):** When a document is first submitted by a project engineering designer to a consultant or client for review and comment, it is called an IFC.
c) **Issue for Approval (IFA):** The consultant usually provides his/her comments on the submitted document using a comment sheet. The project

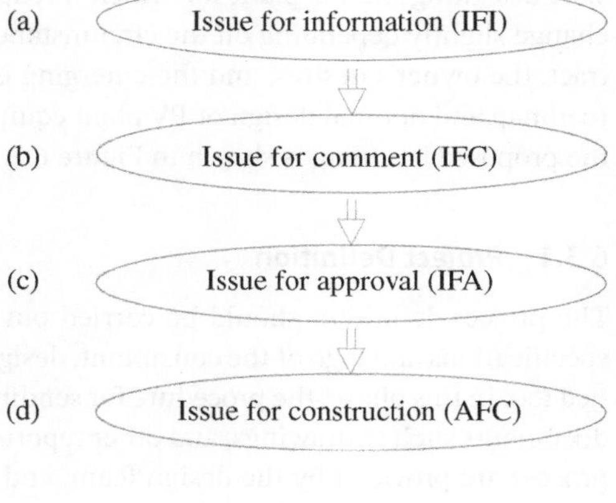
(a)
Issue for information (IFI)
(b)
Issue for comment (IFC)
(c)
Issue for approval (IFA)
(d)
Issue for construction (AFC)

Figure 6.6 Terminologies commonly used in submitted version of engineering documents.

designer submits the revised document for final approval to the consultant. This document is called IFA.

d) **Approved for Construction (AFC):** An engineering certificate, which is approved by the consultant, is sent to the contractors or the construction department under the name of AFC certificate.

After the final drawings are sent to the contractor, he/she starts the execution operation. When each zone is completed, as-built maps are prepared. The as-built map is the final draft of engineering documents, which is archived for the LS-PVPP and must be signed by the project pillars. The as-built maps are prepared for two purposes:

1) To check if there are any mistakes in the position of equipment. The position of one or more equipment may have changed due to human error during the execution.
2) To get final approval from the owner for a change in the position of an equipment. Such change may be needed due to an unforeseen problem, which is identified during the execution.

6.3 Roadmap Proposal for LS-PVPP Design

LS-PVPPs are installed in different geographical locations and climates, where each PV plant has its own unique characteristics. Designers and engineers in every project face new and unpredictable issues and challenges. For this reason, the design steps for all LS-PVPPs are not the same, and engineers must consider different solutions to the obstacles and issues specific to each project. Solutions should be determined such that the operation of PV power plant does not face technical problems and the return on project capital is within the acceptable limits.

The breadth of an LS-PVPP design document allows the engineering team to start designing the PV plant for which a roadmap is needed. The roadmap may change slightly depending on the circumstances of each project, the type of contract, the owner's desires, and the emerging issues. In this section, a method for roadmap and optimal design of PV plant equipment is introduced. A flowchart of the proposed roadmap is shown in Figure 6.7.

6.3.1 Project Definition

The project definition should be carried out in the first step. The project title, specifications and logo of the consultant, design team, and owner should be specified too. In this phase, the procedure for sending documents, the format of various documents such as drawings, and other reports are specified. The design steps and process are provided by the design team, and the stakeholders of the project are

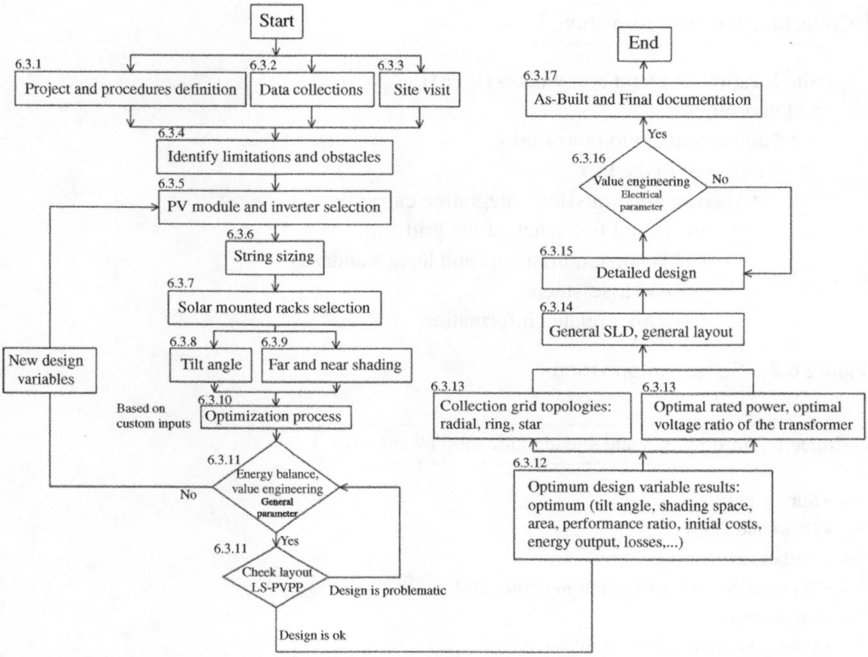

Start
6.3.1
Project and procedures definition
6.3.2
Data collections
6.3.3
Site visit
6.3.4
Identify limitations and obstacles
6.3.5
PV module and inverter selection
6.3.6
String sizing
6.3.7
Solar mounted racks selection
6.3.8
Tilt angle
6.3.9
Far and near shading
New design variables
Based on custom inputs
6.3.10
Optimization process
6.3.11
Energy balance, value engineering
General parameter
No
Yes
6.3.11
Cheek layout LS-PVPP
Design is problematic
Design is ok
6.3.12
Optimum design variable results: optimum (tilt angle, shading space, area, performance ratio, initial costs, energy output, losses,...)
6.3.13
Collection grid topologies: radial, ring, star
6.3.13
Optimal rated power, optimal voltage ratio of the transformer
6.3.14
General SLD, general layout
6.3.15
Detailed design
6.3.16
Value engineering
Electrical parameter
No
Yes
6.3.17
As-Built and Final documentation
End

Figure 6.7 A flowchart of the proposed roadmap for optimal design of PV plant. *Source:* Şenol et al. [3] and Chen et al. [4].

informed about the design process from the beginning so that in case of any dispute over the design process, the issue can be resolved faster.

6.3.2 Collecting General Information

One person as the head or manager of the design team depending on the expertise and scope of work collects the basic information about the project. The headings below are the important information to be collected as shown in Figure 6.8.

The information shown in Figure 6.8 can be obtained from satellite maps, network connection reports, and/or feasibility reports.

6.3.3 Collecting Information By Site Visit

The engineering team should not rely only on the satellite maps and the previous maps. They should visit the site several times and take note of the existing executive and design barriers. The mapping should be performed using advanced engineering tools and by the flying robot so that its output includes topographic maps, digital elevation model (DEM), Digital Surface Model (DSM), and orthomosaic

Collecting general information

- Site location and land coordinates (UTM)
- Meteorology
- Site assessment to power grid
- Distance to the PCC
- Maximum permissible integration capacity
- Voltage and frequency of the grid
- Grid code requirements and local standards
- Road assessment
- Other related information

Figure 6.8 General information.

Collecting information and site identification by site visit

- Survey plan
- DEM and DSM
- Orthomosaic image
- Topography and contours from terrain grid
- Slope analysis
- Land geotechnics and seismicity
- Drainage, seasonal flooding, and watersheds
- Geotechnical surveying and soil testing
- Other related information
- Ramming and pull out tests
- Land cover
- Far, near shading and other site conditions
- Critical challenges and barrier
- Air pollution and suspended solid particles

Figure 6.9 Collecting information by site visit.

map of the site. The following information are obtained during the site visit as shown in Figure 6.9.

6.3.4 Limitations and Obstacles Identification

After reviewing the information collected in Sections 6.3.2 and 6.3.3, the limitations and obstacles of implementation are identified. The limitations and obstacles may include the following:

1) Physical constraints: topography, percentage of land profile slope, waterways inside and around the site, and internal access roads

2) Executive restrictions and problems caused by soil condition
3) Limitations of land area, shading of hills and mountains around the site
4) Constraints such as the shape of land, which leads to an increase in cable and other losses
5) Other restrictions

These limitations should be analyzed in the next steps after identification. The project coordinator sends the information discussed in Sections 6.3.1–6.3.4 to the specialists with corresponding expertise.

6.3.5 PV Module and Inverter Selection

The basic technical specifications of PV module and inverter are selected according to the site conditions and the limitations discussed in Section 6.3.4. In Chapter 7, the selection criteria for PV module and inverter will be discussed in detail.

6.3.6 String Size Calculations

Using the basic specifications of PV modules and inverter, the specifications of the strings are obtained. The following criteria must be considered for further examination in the next steps of the design:

- The allowed number of PV modules in series for each string
- The allowed number of strings for each inverter
- The allowed number of inverters

The calculations are performed in order to verify the optimal size and flexibility of the block design and to come up with an optimal design layout.

6.3.7 Solar PV Mounting Structure Selection

The choice of initial design of the structure depends on the type of power plant installation. Generally, LS-PVPP can be either ground mounted or floating mounted. The process and methodology of solar PV mounting structure selection is presented in Figure 6.10.

By examining the indicators of Figure 6.10 and the executive constraints, the initial model of the structure is selected and modeled by structural analysis software. Based on the optimal angle limits from the history of previous projects and latitude, mechanical calculations of the structure are performed. After determining the optimal angle as will be discussed in Section 6.3.11, the detailed drawings of the PV power plant are prepared.

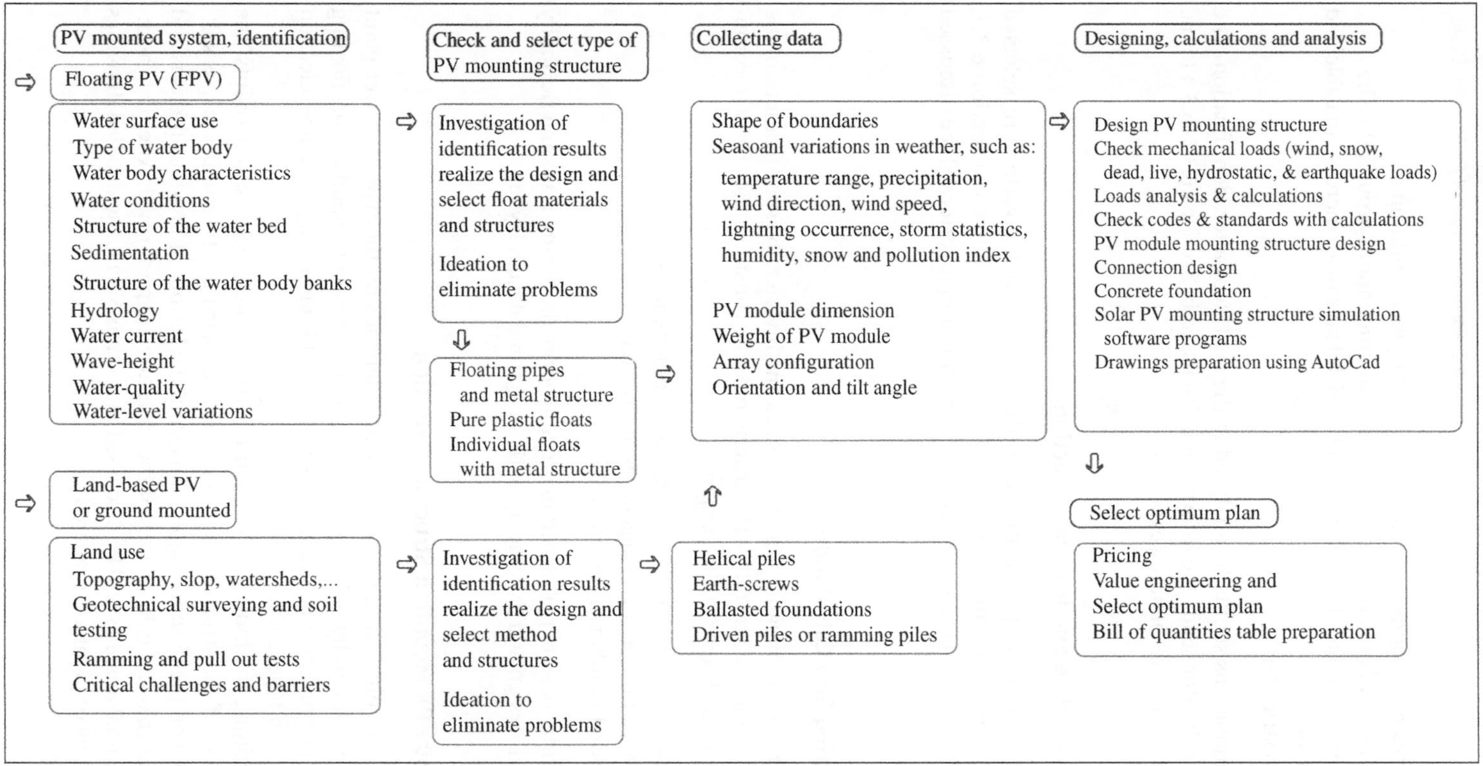
PV mounted system, identification
Floating PV (FPV)
Water surface use
Type of water body
Water body characteristics
Water conditions
Structure of the water bed
Sedimentation
Structure of the water body banks
Hydrology
Water current
Wave-height
Water-quality
Water-level variations
Land-based PV
or ground mounted
Land use
Topography, slop, watersheds,...
Geotechnical surveying and soil
testing
Ramming and pull out tests
Critical challenges and barriers
Check and select type of
PV mounting structure
Investigation of
identification results
realize the design and
select float materials
and structures
Ideation to
eliminate problems
Floating pipes
and metal structure
Pure plastic floats
Individual floats
with metal structure
Investigation of
identification results
realize the design and
select method
and structures
Ideation to
eliminate problems
Collecting data
Shape of boundaries
Seasoanl variations in weather, such as:
temperature range, precipitation,
wind direction, wind speed,
lightning occurrence, storm statistics,
humidity, snow and pollution index
PV module dimension
Weight of PV module
Array configuration
Orientation and tilt angle
Helical piles
Earth-screws
Ballasted foundations
Driven piles or ramming piles
Designing, calculations and analysis
Design PV mounting structure
Check mechanical loads (wind, snow,
dead, live, hydrostatic, & earthquake loads)
Loads analysis & calculations
Check codes & standards with calculations
PV module mounting structure design
Connection design
Concrete foundation
Solar PV mounting structure simulation
software programs
Drawings preparation using AutoCad
Select optimum plan
Pricing
Value engineering and
Select optimum plan
Bill of quantities table preparation

Figure 6.10 The methodology of solar PV mounting structure selection. *Source:* Energy Sector Management Assistance Program and Solar Energy Research Institute of Singapore [5].

6.3.8 Tilt Angle Calculation

By determining the number of modules, string size, and the dimensions and type of PV module arrangement on the structure, the monthly, seasonal, and annual slope angles are calculated. Slope angle can be determined from two methods presented in references [6–8]. An approach is discussed in Section 6.3.11 to obtain the optimal values of the slope angles in order to achieve the maximum radiation on the module surface.

6.3.9 Calculations of Far and Near Shading

In Step 1, for the selected structure and the desired angles, shading intervals are calculated according to [9]. The followings should be considered when calculating the shading interval.

- Shading distance between rows should be calculated for December 21.
- Shading distance between rows should be calculated for several angles.
- Percentage of shading losses for each angle should be calculated and recorded.

In Step 2, dynamic hill shading analysis is performed using the site DEM and its surrounding area. This analysis is carried out using ArcGis, global mapper, or similar software. Using the dynamic hill shading analysis, the shadows of the hills and heights around and inside the site are determined by changing the solar path. Furthermore, the critical points with remote shading are identified. Figure 6.11 shows a typical image of the dynamic hill shading analysis. Using the map shown in Figure 6.11, the location constraints in terms of far or near shading on the site are determined, and an alternative solution is proposed [10].

In Step 3, the percentage of far and near shading losses is calculated. For further explanation, the reader is referred to [11]. The results of Steps 1–3 are used as inputs to prepare the energy balance report.

6.3.10 Optimization Process

The optimization process is generally defined based on the constraints and obstacles for constructing the PV plant. The important criteria for optimizing PV plant are shown in Figure 6.12. The optimization process can be examined using one or more criteria. The plant Total Energy and Total Cost are the most widely used optimization indicators for PV power plants [12].

The optimization process is carried out using the indicators below, which are also shown in Figure 6.12 [12].

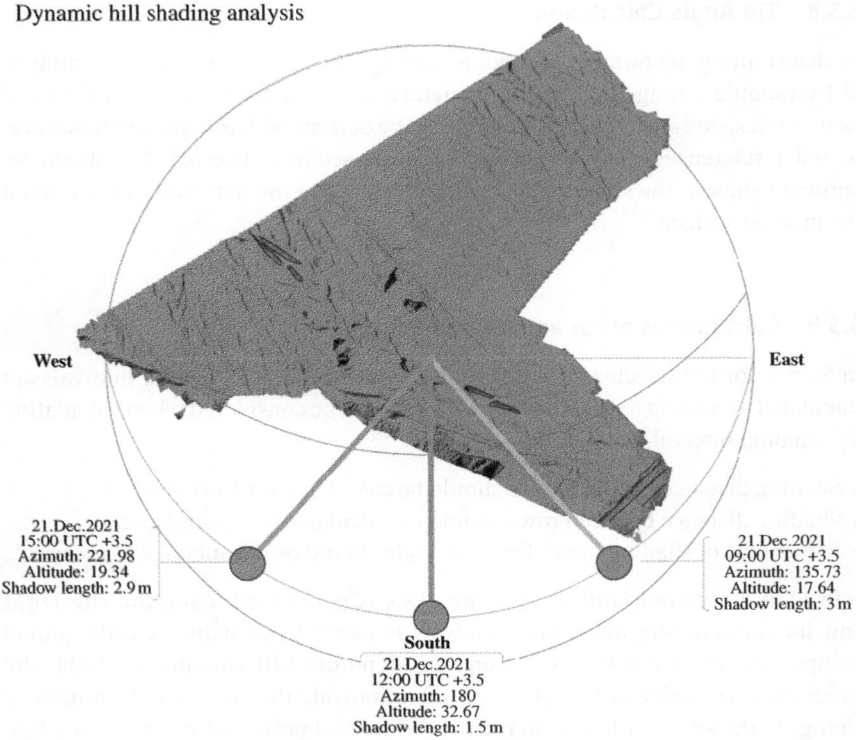
Dynamic hill shading analysis
West
East
21.Dec.2021
15:00 UTC +3.5
Azimuth: 221.98
Altitude: 19.34
Shadow length: 2.9 m
21.Dec.2021
09:00 UTC +3.5
Azimuth: 135.73
Altitude: 17.64
Shadow length: 3 m
South
21.Dec.2021
12:00 UTC +3.5
Azimuth: 180
Altitude: 32.67
Shadow length: 1.5 m

Figure 6.11 The typical image of the dynamic hill shading analysis. *Source:* Horn [10].

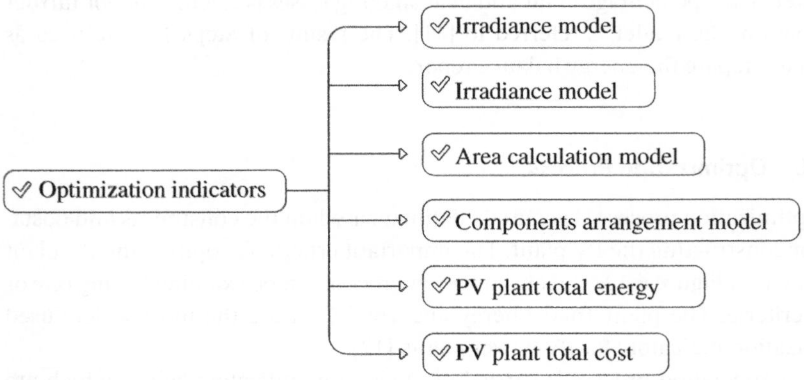
Optimization indicators
Irradiance model
Irradiance model
Area calculation model
Components arrangement model
PV plant total energy
PV plant total cost

Figure 6.12 Important criteria and indicators for optimizing PV plant. *Source:* Zidane et al. [12].

Irradiance Model: In this model, it is examined that which of the following factors provides the most radiation on the PV modules.

- Slope angle
- Slope direction of the structure toward east–west or south
- Cracked or fixed structure
- Seasonal or annual type for the slope angle of fixed structure
- Bifacial modules
- White stones on the floor of PV plant to increase light reflection

Area Calculation Model: This model is employed for the cases where the installation land of PV power plant is limited and/or the land does not have a uniform shape. The optimization process is carried out based on shading and placement of the tables. It is possible to install the PV module with the highest capacity in a minimum space.

Components Arrangement Model: Based on this model, the optimization process is performed by selecting the best arrangement of PV modules, blocks, and arrays.

PV Plant Total Energy: The optimization process is based on the total annual energy obtained from PV power plant. Various designs are analyzed, and the one with the highest total annual energy is selected.

PV Plant Total Cost: The optimization process is based on fixed and operating costs of the project. Among various designs, the one with more favorable economic indicators is selected as the optimal design.

6.3.11 Energy Balance and Value Engineering

By selecting an optimization process, general parameters of the PV power plant are optimized. Moreover, based on the value engineering, the optimal design is carried out. The general parameters include those affecting layout of the PV plant such as module and inverter capacity, structure dimensions and type, slope angle, near and far shadings, row shading distance, ground slope, and other elements in the blocks layout.

In optimizing general parameters, the following points should be considered:

1) The optimization process depends on the owner's choices and/or the special conditions governing the projects.
2) Since the optimized results are unique to every specific process, only one optimization approach should be selected.

Example 1: Tilt angle optimization

When the irradiance model is selected, a slope angle of 30° is the optimal annual angle. If the PV plant total energy method is selected for optimization, a slope angle of 25° is the optimal angle. Since the table height for 25° is less than that of 30°, the shading and losses, and radiation at 25° are less than those of 30°. As a result, the annual production capacity of the PV plant at the angle of 25° increases. It is to be noted that the initial cost of structure for a 25° angle is less than that of 30°.

Example 2: Shading distance optimization
If the area calculation model is selected, the shadow spacing of rows should be as small as possible to create more space for PV installation. If the PV plant total energy method is chosen, the shadow spacing of the rows should be increased to the extent that the PV plant production is maximized by reducing the shading losses.

3) Optimizing the PV power plant should not reduce the safety requirements of the project or violate the standards. For example, when optimizing the structure design, the weight of structure holding the modules cannot be reduced without performing mechanical calculations and considering a proper reliability factor. In case of strong winds, the structure and modules can be moved and the safety inside and around the site may be jeopardized.
4) After optimizing the general parameters, the layout of PV power plant should be reexamined. If the design is acceptable, the optimal values are final. Otherwise, the energy balance and value engineering must be investigated again.

6.3.12 Optimal Transformer Size

To obtain optimal transformer size, first the location of the high-voltage (HV) substation of the PV power plant is determined. Then, the configuration and topology of the internal network of the plant are investigated, i.e. in terms of ring, radius, and star, and based on the application, an appropriate configuration is chosen. Finally, the optimal transformer size in terms of power and voltage ratio is determined.

6.3.13 General SLD and Layout

An SLD is prepared based on the technical specifications and number of PV modules, inverters, strings, transformers, and other principal equipment. The details of the SLD such as cable size and protections are calculated and included in the final document.

The location map is designed with the provided specifications as general optimal parameters. In the location map, PV modules, structures, inverters, transformers, row shading distances, switches, and other equipment are determined. The detailed maps such as wiring, SCADA, CCTV, and ground system are prepared and included to the final map.

6.3.14 Detailed Design

The procedures discussed in Sections 6.3.5–6.3.13 are repeated several times to come up with the best option. After preparing the location map and single-line diagram, the detailed design documents are produced based on the list of MDL engineering documents and the project schedule. The detailed design documents provide technical specifications of all cables, protection system, ground system, monitoring system, CCTV, CT and PT devices, output feeders, wirings, DC and AC panels, foundations, facility building, and other equipment.

6.3.15 Electrical Parameters and Value Engineering

After calculating the electrical parameters affecting the cost and revenue of a PV power plant, value engineering is performed. Based on the value engineering, a design that incurs less costs and provides higher productivity and revenue is selected. Important electrical parameters involving the value engineering include DC cables, AC cables, number of DC panels, and number of AC panels.

6.3.16 Preparing Final Documents

This is the last stage of design and engineering services to be performed before constructing PV plant. When the documents discussed in Section 6.3.15 are produced, there may exist other challenges and issues. That should be overcome by discussions among the engineering team, consultants, and the owner. Following the discussions and once the final modifications are made, the documents will be approved as final or as-built documents.

6.4 Conclusion

The design of an LS-PVPP involves different phases. In each project, depending on the climate and geography, new and unpredictable problems arise, which makes the design of the power plant more complex. Many documents are prepared in the design process. They include calculations, maps, and reports.

Based on the example cases presented in this chapter, it is concluded that the LS-PVPP design process is performed several times for some projects to finally select an optimal design in terms of engineering value.

The concepts presented in this chapter help beginners and interested individuals become familiar with the design process of an LS-PVPP. For further information, the reader can review various articles and reports to acquire the required experience and knowledge about the design of LS-PVPP.

References

1 Wulff, I.A., Rasmussen, B., and Westgaard, R.H. (2000). Documentation in large-scale engineering design: information processing and defensive mechanisms to generate information overload. *International Journal of Industrial Ergonomics* 25 (3): 295–310.

2 Anderson, P.O., Knoben, J.E., and Troutman, W.G. (2003). *Handbook of Electrical Engineering-For Practitioners in the Oil, Gas and Petrochemical Industry*. John Wiley & Sons.

3 Şenol, M., Abbasoğlu, S., Kükrer, O., and Babatunde, A.A. (2016). A guide in installing large-scale PV power plant for self consumption mechanism. *Solar Energy* 132: 518–537.

4 Chen, S., Li, P., Brady, D., and Lehman, B. (2013). Determining the optimum grid-connected photovoltaic inverter size. *Solar Energy* 87: 96–116.

5 Energy Sector Management Assistance Program and Solar Energy Research Institute of Singapore (2019). *Where Sun Meets Water: Floating Solar Handbook for Practitioners*. Washington, DC: World Bank.

6 Ghosh, H.R., Bhowmik, N.C., and Hussain, M. (2010). Determining seasonal optimum tilt angles, solar radiations on variously oriented, single and double axis tracking surfaces at Dhaka. *Renewable Energy* 35 (6): 1292–1297.

7 Abdallah, R., Juaidi, A., Abdel-Fattah, S., and Manzano-Agugliaro, F. (2020). Estimating the optimum tilt angles for south-facing surfaces in Palestine. *Energies* 13 (3): 623.

8 Sayigh, A.A.M. (ed.) (2012). *Solar Energy Engineering*, 60. Elsevier.

9 International Electrotechnical Commission (2015). *IEC 61724-1 Photovoltaic System Performance–Part 1: Monitoring (New Work in Process)*. Geneva, Switzerland: IEC Central Office.

10 Horn, B.K. (1981). Hill shading and the reflectance map. *Proceedings of the IEEE* 69 (1): 14–47.

11 Mermoud, A. (2010). *Modeling Systems Losses in PVsyst*. Institute of the Environmental Sciences Group of Energy–PVsyst: Universitè de Genève.

12 Zidane, T.E.K., Adzman, M.R., Tajuddin, M.F.N. et al. (2020). Optimal design of photovoltaic power plant using hybrid optimisation: a case of south Algeria. *Energies* 13 (11): 2776.

7

PV Power Plant DC Side Design

7.1 Introduction

In general, the electrical part of a photovoltaic (PV) power plant from PV modules to inverter input is called the plant DC side. It is the largest part of a PV plant in terms of funding and land area. Therefore, it is important to come up with an optimal design for the plant DC side equipment.

In this chapter, the main components of DC side and the corresponding design methods are presented. More specifically, it is discussed how to design main equipment of the DC side of a large-scale PV power plant (LS-PVPP).

7.2 DC Side Design Methodology

Figure 7.1 shows the design methodology of DC equipment of a PV plant. The plant site information is received as input, and the main equipment is designed by considering technical and financial criteria. According to the methodology of Figure 7.1, to design the main equipment of DC side, it is necessary to understand its functionality details.

Figure 7.2 shows the single-line diagram of the DC side of a PV plant including its main electrical equipment. The main electrical equipment includes PV modules, inverters, connectors, combiner box (also called junction box), DC protections, DC switches, and DC cables.

The most important technical specifications that must be determined for the DC side equipment are as follows:

- PV modules technology
- Inverter configuration
- PV plant size

Step-by-Step Design of Large-Scale Photovoltaic Power Plants, First Edition. Davood Naghaviha, Hassan Nikkhajoei, and Houshang Karimi.

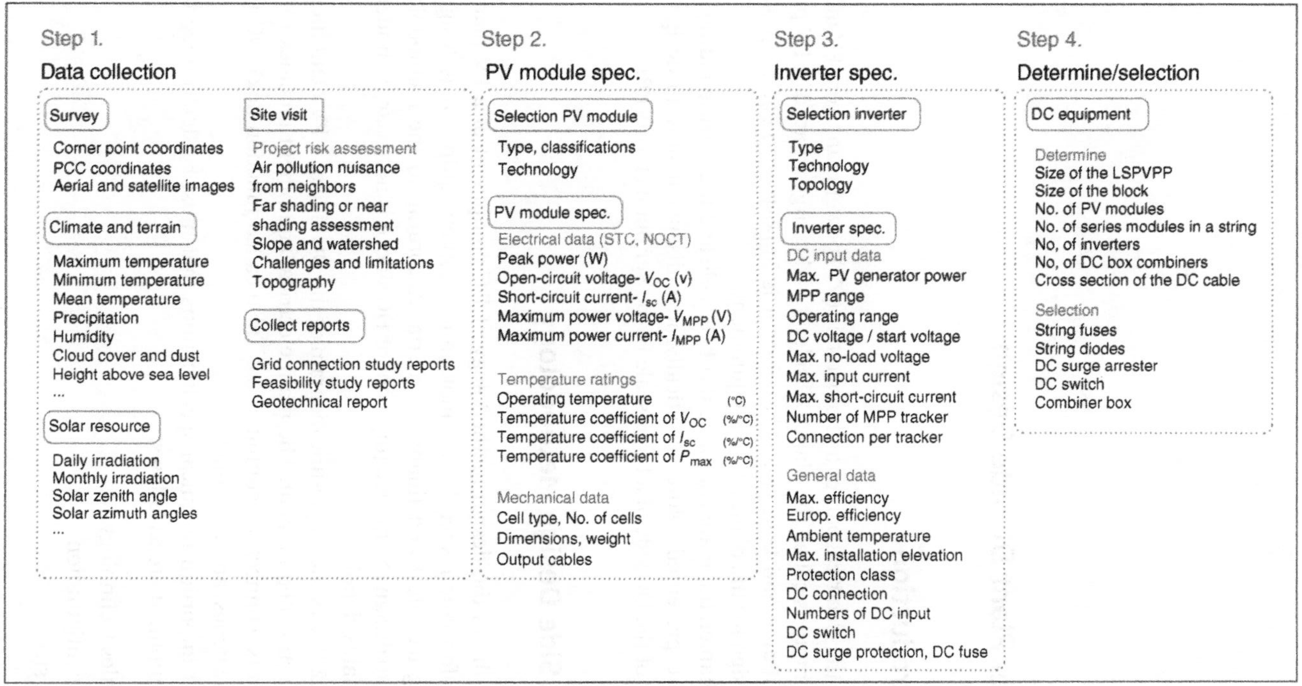
Step 1.
Data collection
Survey
Corner point coordinates
PCC coordinates
Aerial and satellite images
Climate and terrain
Maximum temperature
Minimum temperature
Mean temperature
Precipitation
Humidity
Cloud cover and dust
Height above sea level
...
Solar resource
Daily irradiation
Monthly irradiation
Solar zenith angle
Solar azimuth angles
...
Site visit
Project risk assessment
Air pollution nuisance from neighbors
Far shading or near shading assessment
Slope and watershed
Challenges and limitations
Topography
Collect reports
Grid connection study reports
Feasibility study reports
Geotechnical report
Step 2.
PV module spec.
Selection PV module
Type, classifications
Technology
PV module spec.
Electrical data (STC, NOCT)
Peak power (W)
Open-circuit voltage- V_{OC} (v)
Short-circuit current- I_{sc} (A)
Maximum power voltage- V_{MPP} (V)
Maximum power current- I_{MPP} (A)
Temperature ratings
Operating temperature (°C)
Temperature coefficient of V_{OC} (%/°C)
Temperature coefficient of I_{sc} (%/°C)
Temperature coefficient of P_{max} (%/°C)
Mechanical data
Cell type, No. of cells
Dimensions, weight
Output cables
Step 3.
Inverter spec.
Selection inverter
Type
Technology
Topology
Inverter spec.
DC input data
Max. PV generator power
MPP range
Operating range
DC voltage / start voltage
Max. no-load voltage
Max. input current
Max. short-circuit current
Number of MPP tracker
Connection per tracker
General data
Max. efficiency
Europ. efficiency
Ambient temperature
Max. installation elevation
Protection class
DC connection
Numbers of DC input
DC switch
DC surge protection, DC fuse
Step 4.
Determine/selection
DC equipment
Determine
Size of the LSPVPP
Size of the block
No. of PV modules
No. of series modules in a string
No, of inverters
No, of DC box combiners
Cross section of the DC cable
Selection
String fuses
String diodes
DC surge arrester
DC switch
Combiner box

Figure 7.1 Design methodology of DC equipment of a PV plant.

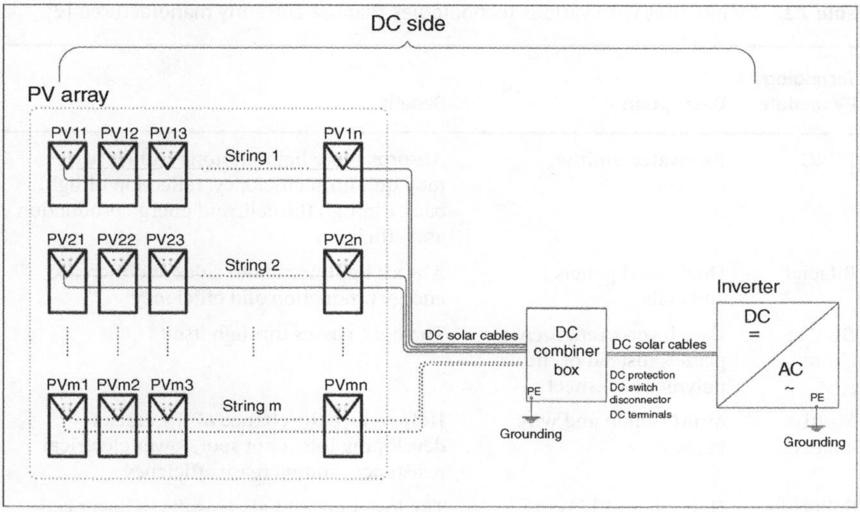
DC side
PV array
PV11 PV12 PV13
PV1n
String 1
PV21 PV22 PV23
PV2n
String 2
PVm1 PVm2 PVm3
PVmn
String m
DC solar cables
DC
combiner
box
PE
Grounding
DC solar cables
DC protection
DC switch
disconnector
DC terminals
Inverter
DC
=
AC
~
PE
Grounding

Figure 7.2 Single-line diagram of DC side of a PV plant. *Source:* Based on Falvo and Capparella [1].

- PV modules number
- Number of series modules in a string
- Number of inverters
- Number of DC combiner box
- DC cable cross section
- Fuses and string diodes
- DC surge arresters
- DC switches

In the following sections, the details on how to design DC side equipment of a PV plant are discussed.

7.3 PV Modules Selection

7.3.1 Module Technology

The classification of various types of PV modules was presented in Chapter 2. Most manufacturers of PV modules offer a wide range of models, including monocrystalline, polycrystalline, and thin films with various output power ranges. In the past few years, there have been many advances in the PV cell technology. Table 7.1 presents PV modules with various technologies that are currently manufactured.

Table 7.1 PV modules with various technologies that are currently manufactured [2].

Technology PV module	Description	Benefit
PERC	Passivated emitter rear cell	Absorbs more light photons and increases total quantum efficiency, reflection of light back through the cell, and energy production and efficiency
Bifacial	Dual-sided panels and cells	Absorb light from both sides and increase energy production and efficiency
Double glass	Uses layer-strengthened glasses instead of the polymer backsheet	The light passes through itself
Multi busbar	Multi ribbon and wire busbars	Help reduce the chance of the cracks developing into a hot spot, lower electrical resistance, and increase efficiency
Split cells	Half-cut and 1/3 cut cells	The lower current also results in lower cell operating temperatures, less busbar shading losses and increased efficiency, and reduced resistive losses
Shingled cells	Overlapping cells	Lower ohmic losses, better area utilization, lower processing temperature, and lower operating temperature resulting in enhanced energy yield
High-density cells	Removing standard vertical 2–3 mm inter-cell gaps	Boost PV module efficiency and reduce area
IBC	Interdigitated back contact cells	Help reduce the chance of cracks developing into a hot spot, lower electrical resistance, and increase efficiency
HJT	Hetero-junction cells	Reduce loses and increase cell efficiency and improved high-temperature performance

Source: Based on Udayakumar et al. [2].

Some manufacturers improve the efficiency and performance of PV modules by integrating several technologies as given in Table 7.1, where each technology has certain applications. For example, for greenhouses, the double glass technology is used as it allows the passage of sunlight. For PV plant sites where there are many birds, the multi busbar technology, interdigitated back contact (IBC) and split cells are suitable. Bird droppings on the surface of a module lead to hot spots. The hetero-junction technology (HJT) is suitable for PV plants where the temperature is extremely high. For PV plants with space limitation, the high-density cells and shingled cells technologies are usually employed. In snowy areas, lakes, or areas with white pebbles, modules with bifacial technology is suitable. For such technology, the reflection of sunlight from white surfaces results in high performance.

7.3.2 PV Module Size

Today, PV modules are produced in a variety of power and voltage ratings. In LS-PVPPs, the module size is generally from 300 to 500 W. The advantages of PV modules with higher output power ratings, e.g. 500 W, include less space needed for the placement, reduced installation costs due to reduced number of modules, reduced cost of structure, and lower cost for DC cabling. The disadvantages of such high power modules, however, are transportation and installation issues due to the large size of the module, high power loss in case of defect or failure of the module, and difficulties in the modules washing operation.

7.3.3 Selection Criteria

The main equipment of a PV power plant is the PV module. There are numerous parameters to consider when selecting PV modules/solar panels. There are more than 4000 solar panel manufacturers in the global market, most of which are located in China. Various types of solar panels are offered by these manufacturers in the global market, and each type of panel has its own advantages and disadvantages. For a PV power plant, depending on the environmental and financial conditions, general scenarios of selecting PV modules are considered.

Figure 7.3 shows the criteria and steps for selecting the PV modules considering technical and financial constraints. According to this figure, six steps must be followed to select the PV modules. In the first step, the classification of PV module is chosen. In the second step, technology of PV module is chosen, and in the third step, the vendor lists are specified based on the required module specifications. In the fourth step, the technical and economic evaluation of similar products is performed. In the fifth step, the selected PV modules are prioritized. Finally (sixth step), a superior PV module type is selected.

7.4 Inverter Selection

Solar inverters are electronic devices that convert the DC power generated by the PV modules into the AC power. To achieve a high-quality performance for a PV power plant, it is greatly important to choose a suitable inverter type for the plant. The inverter selection methodology is carried out in six steps as shown in Figure 7.4.

In the first step, the inverter type is specified such that it can operate in the grid-connected mode. It should be noted that under normal conditions, a PV plant should be connected to the grid. In the second step, the PV inverter configuration is evaluated in terms of technical requirements, applications, advantages, and disadvantages. Consequently, a superior configuration is selected. In the third step, the inverter vendor list is prepared. In the fourth step, the technical and economic

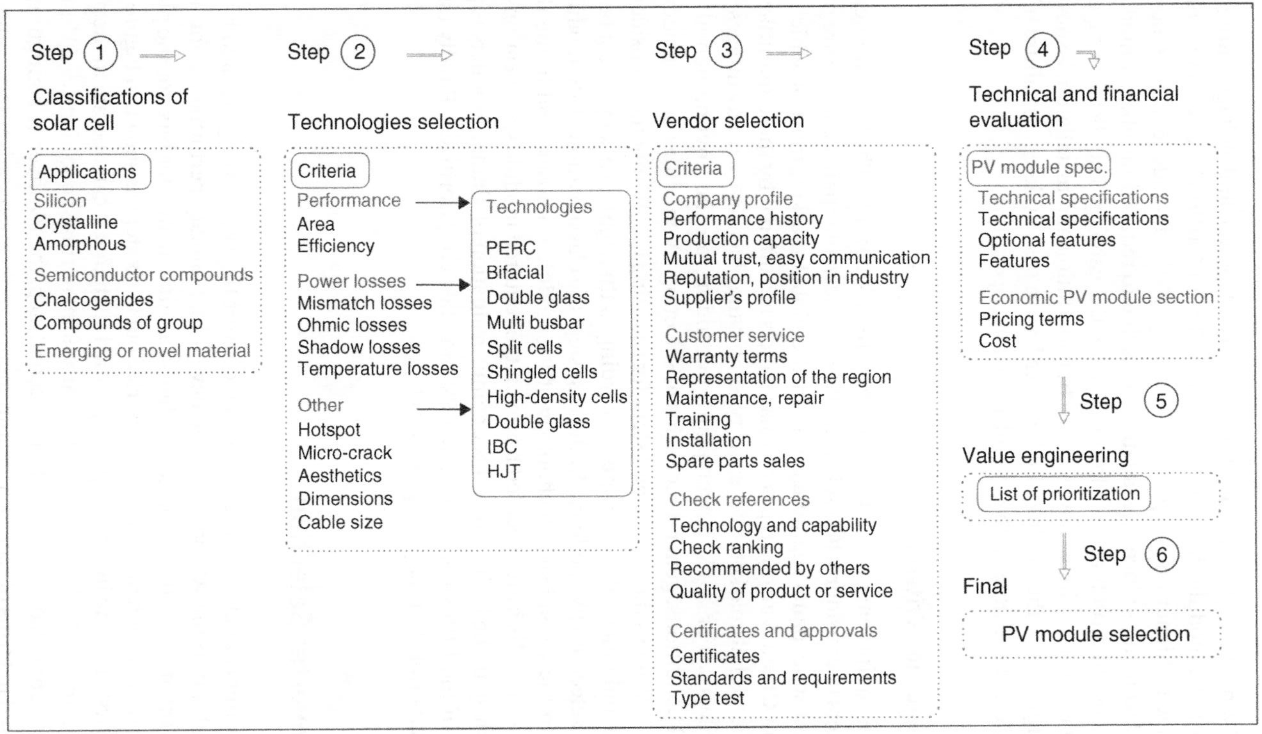

Step 1
Classifications of solar cell
Applications
Silicon
Crystalline
Amorphous
Semiconductor compounds
Chalcogenides
Compounds of group
Emerging or novel material
Step 2
Technologies selection
Criteria
Performance
Area
Efficiency
Power losses
Mismatch losses
Ohmic losses
Shadow losses
Temperature losses
Other
Hotspot
Micro-crack
Aesthetics
Dimensions
Cable size
Technologies
PERC
Bifacial
Double glass
Multi busbar
Split cells
Shingled cells
High-density cells
Double glass
IBC
HJT
Step 3
Vendor selection
Criteria
Company profile
Performance history
Production capacity
Mutual trust, easy communication
Reputation, position in industry
Supplier's profile
Customer service
Warranty terms
Representation of the region
Maintenance, repair
Training
Installation
Spare parts sales
Check references
Technology and capability
Check ranking
Recommended by others
Quality of product or service
Certificates and approvals
Certificates
Standards and requirements
Type test
Step 4
Technical and financial evaluation
PV module spec.
Technical specifications
Technical specifications
Optional features
Features
Economic PV module section
Pricing terms
Cost
Step 5
Value engineering
List of prioritization
Step 6
Final
PV module selection

Figure 7.3 The criteria and steps for selecting the PV modules. *Source:* Based on El-Bayeh et al. [3].

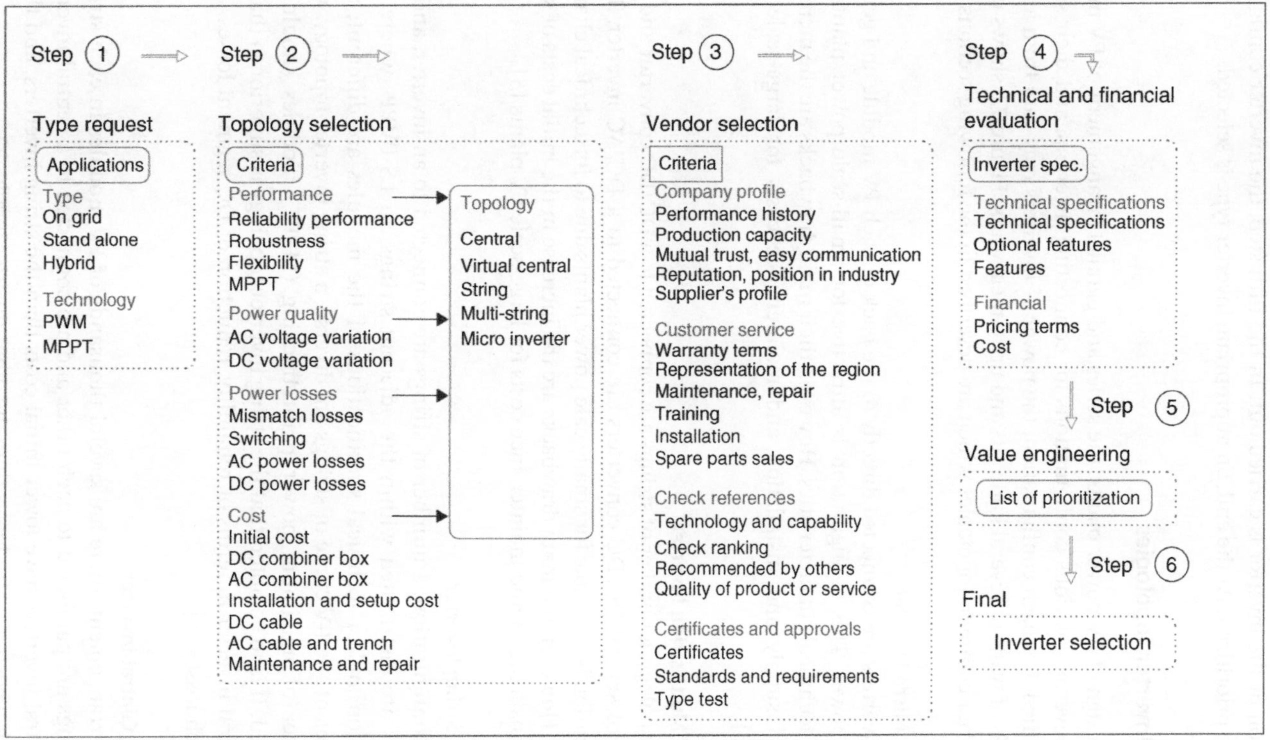
Step 1
Type request
Applications
Type
On grid
Stand alone
Hybrid
Technology
PWM
MPPT
Step 2
Topology selection
Criteria
Performance
Reliability performance
Robustness
Flexibility
MPPT
Power quality
AC voltage variation
DC voltage variation
Power losses
Mismatch losses
Switching
AC power losses
DC power losses
Cost
Initial cost
DC combiner box
AC combiner box
Installation and setup cost
DC cable
AC cable and trench
Maintenance and repair
Topology
Central
Virtual central
String
Multi-string
Micro inverter
Step 3
Vendor selection
Criteria
Company profile
Performance history
Production capacity
Mutual trust, easy communication
Reputation, position in industry
Supplier's profile
Customer service
Warranty terms
Representation of the region
Maintenance, repair
Training
Installation
Spare parts sales
Check references
Technology and capability
Check ranking
Recommended by others
Quality of product or service
Certificates and approvals
Certificates
Standards and requirements
Type test
Step 4
Technical and financial evaluation
Inverter spec.
Technical specifications
Technical specifications
Optional features
Features
Financial
Pricing terms
Cost
Step 5
Value engineering
List of prioritization
Step 6
Final
Inverter selection

Figure 7.4 The criteria and steps for selecting the solar inverter. *Source:* Based on Cabrera Tobar [4].

evaluation of the inverter is carried out. In the fifth step, the inverter configurations are prioritized. At the end, an appropriate inverter type is selected.

7.4.1 Inverter Topologies

In the design of PV power plants, the series and parallel arrangement of PV modules to inverters, various configurations for connecting inverters, and their sizes are specified. For each configuration, the power of inverters may vary from a few hundreds of watts to several kilowatts and up to megawatts. Figure 7.5 shows various inverter common topologies which are described in the following sections [4].

7.4.1.1 Micro Inverter

Micro inverters are connected directly to the back of each PV module and generate AC power. This configuration is attractive for small-scale power plants in terms of technical characteristics. However, their main drawbacks are the increase in initial supply and installation and maintenance costs for large-scale PV plants [5].

7.4.1.2 Multi-string Inverter

In this topology, a number of strings are connected to a DC/DC converter and the outputs of several DC/DC converters are connected to a DC/AC inverter. This configuration is desirable for small-scale power plants due to its technical characteristics. However, the main drawbacks are the increase in the initial costs of supply and installation and maintenance costs for large-scale PV plants [5].

7.4.1.3 String Inverter

In this configuration, a number of strings are connected to an inverter, and all inverters are distributed within the adjacent strings. In LS-PVPPs where the orientation and/or technical specifications of the modules are different, and variations of the I-V curve of strings are diverse, a string inverter topology with a number of maximum power point tracking (MPPT) modules should be employed. It is to be noted that the string inverter topology is superior to that of the central inverter in terms of reliability, reduction of DC segment losses, and mismatch losses [5].

7.4.1.4 Central Inverter

In this arrangement, there are several thousands of PV modules in each string. The strings are paralleled to each other and are connected to a central inverter. The central inverters have lower initial costs than the string inverters, and their installation costs are also lower than the other inverter topologies.

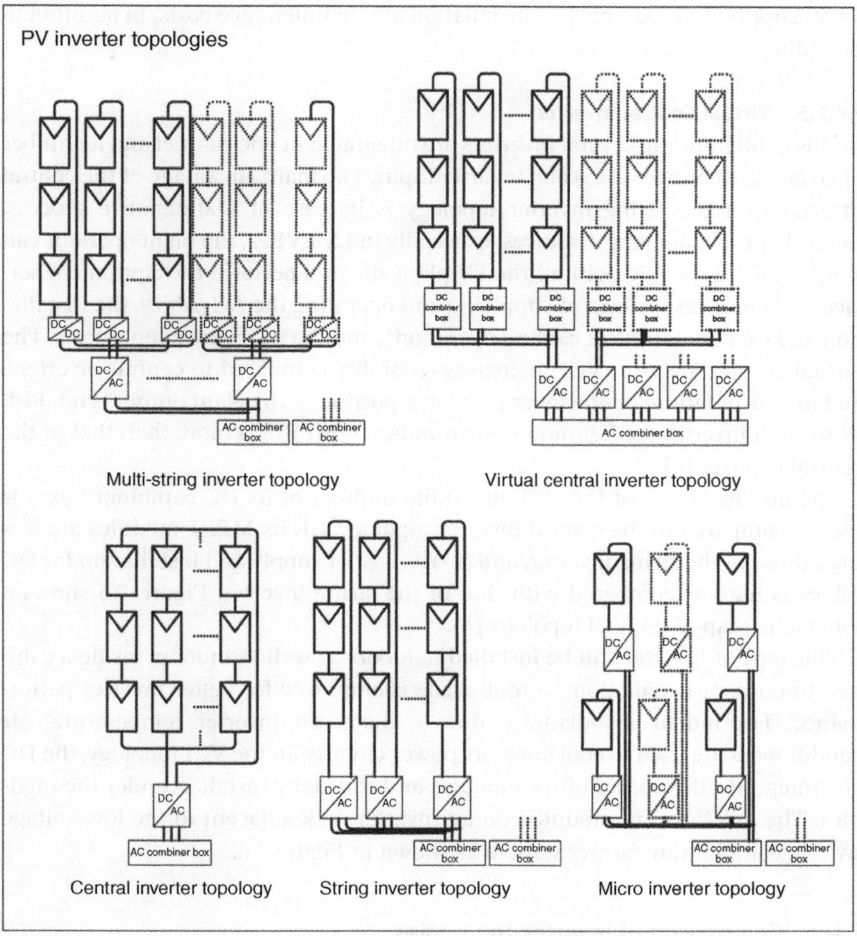
PV inverter topologies
DC DC
DC AC
AC combiner box
Multi-string inverter topology
DC combiner box
Virtual central inverter topology
Central inverter topology
String inverter topology
Micro inverter topology

Figure 7.5 Various inverter topologies. *Source:* Based on Cabrera Tobar [4].

The main problems of central inverters are the increase in costs related to DC cables and reduced reliability in case of failure of one of the central inverters where a large portion of the plant power production is lost. The mismatch losses of this inverter configuration are higher than the other types. Moreover, tracking of the maximum power point is more complicated to achieve compared to the other topology due to its more advanced technology than other topologies [5]. It is to be noted that the central inverter topology is much more welcomed by the engineering, procurement, and construction (EPC) contractor team since it needs lower supply, and requires less installation and maintenance costs, in addition to providing higher reliability.

7.4.1.5 Virtual Central Inverter

In this configuration, several inverters are integrated in one cabinet and a number of strings are connected to each inverter input. The main advantage of the central inverter over the string inverter topology is its ease of maintenance process. Instead of traveling long distances, especially in LS-PVPPs, the plant operator can simply go to a specific point of the PV plant site and perform the equipment service. This reduces the cost of supplying and operating the AC cables, the installation and execution time of cabling operations, and the AC boxes supply costs. The virtual central inverter (VCI) increases reliability compared to central inverters, and in case of failure of one inverter, a large portion of the plant output is not lost. With such inverter configuration, the number of MPPTs is more than that of the central inverter [6].

The disadvantages of the VCI are (i) the number of its DC combiner boxes is higher compared to the central inverter topology, (ii) its MPPT modules are less than those of the string inverter, and (iii) its cost of supply and installation for DC cables is high as compared with that of the string inverter. Figure 7.6 shows a schematic map of the VCI topology [6].

This type of inverter can be installed outdoors as well as indoors inside a cabinet. In outdoor installations, a roof shade is employed for better inverter performance. For indoor installations, due to increased inverter temperature, air conditioners are used to cool down its power circuits. In the VCI topology, the DC box integrates the output of PV modules and is usually installed under the modules. The inverters are mounted on an inverter rack adjacent to the low-voltage switchgear and transformer station, as shown in Figure 7.6.

7.4.2 Comparison of Inverter Topologies

In Table 7.2, various inverter topologies are compared with respect to their applications and technical and financial conditions. The topologies are compared in terms of performance, power quality, power loss and cost. The quality indicators are 0 for very low, 1 for low, 2 for medium, 3 for high, and 4 for very high. Details of comparison criteria are described below:

- **Performance:** considers the strength, reliability, flexibility, and performance of MPPT module;
- **Power quality:** indicates the quality of DC and AC voltages;
- **Power loss:** includes mismatch losses, switching, and AC and DC losses;
- **Costs:** include initial cost as well as installation and maintenance expenses.

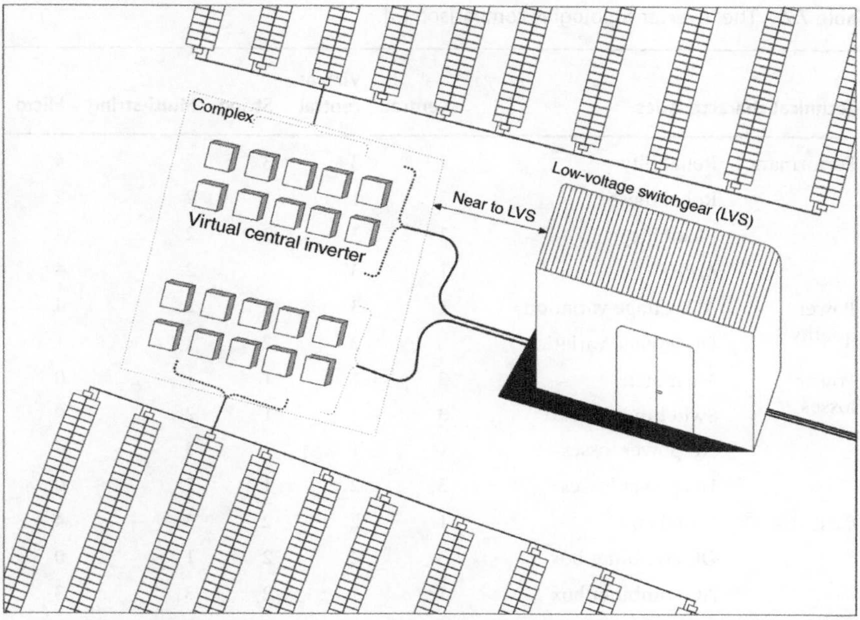
Complex
Virtual central inverter
Near to LVS
Low-voltage switchgear (LVS)

Figure 7.6 A schematic drawing of the VCI configuration. *Source:* Based Desimpelaere and Schimpf [6].

From Table 7.2, the virtual central configuration for LS-PVPP has a higher priority in terms of technical and economic aspects than the central inverter and string inverter topologies.

7.5 PV Modules Number

To determine the number of PV modules, it is necessary to know the following information for each string:

- Technical specifications of PV module
- Technical specifications of the inverter including its operating voltage range and permitted input/output currents
- Maximum and minimum annual site temperature

Figure 7.7 shows the AB diagram and current–voltage (I-V) curve for a PV module for a given radiation. The CD curve shows the power–voltage (P-V) curve. Each PV module has a single operating point called the maximum power point (MPP). When the system is at MPP, it generates maximum output power. The I-V

Table 7.2 The inverter topologies comparison [5].

Technical characteristics		Central	Virtual central	String	Multi-string	Micro
Performance	Reliability	1	3	3	2	4
	Robustness	3	1	1	2	2
	Flexibility	1	3	3	2	4
	MPPT	1	3	3	2	4
Power quality	AC voltage variation	1	3	3	2	4
	DC voltage variation	3	2	2	1	1
Power losses	Mismatch	3	1	1	1	0
	Switching	3	1	1	2	0
	AC power losses	1	1	3	3	4
	DC power losses	3	2	1	2	0
Cost	Initial cost	1	2	2	3	4
	DC combiner box	3	2	2	1	0
	AC combiner box	1	1	2	3	3
	Installation and setup cost	2	2	3	2	4
	DC cable	3	3	1	2	0
	AC cable and trench	2	2	3	2	4
	Maintenance and repair	1	2	3	4	4

Source: Based Kolantla et al. [5].

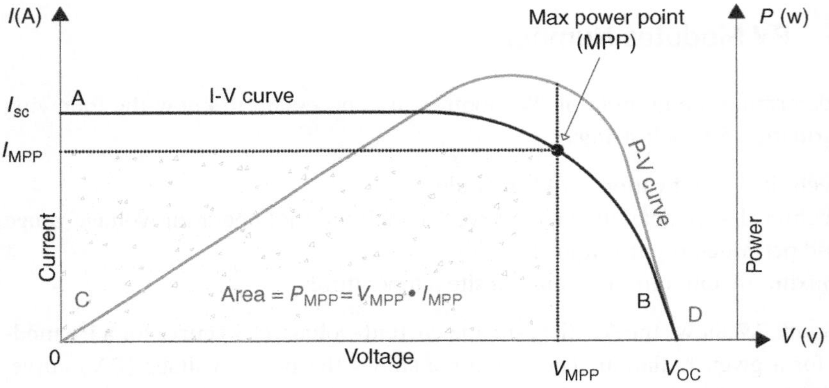
I(A)
Max power point
(MPP)
P (w)
Isc
A
I-V curve
IMPP
P-V curve
Current
Power
C
Area = PMPP= VMPP • IMPP
B
D
V (v)
0
Voltage
VMPP
VOC

Figure 7.7 The I-V curve and P-V curve for a PV module for a given irradiation. *Source:* Based on Spertino et al. [7].

curve of a PV module is nonlinear and time varying and depends on various factors such as irradiation, temperature, and the module's internal resistance.

Figure 7.8 shows the effect of irradiation intensity and temperature on the I-V curve. This figure also shows that:

a) As the surface temperature of a PV module increases, the voltage and output power of the module decreases.

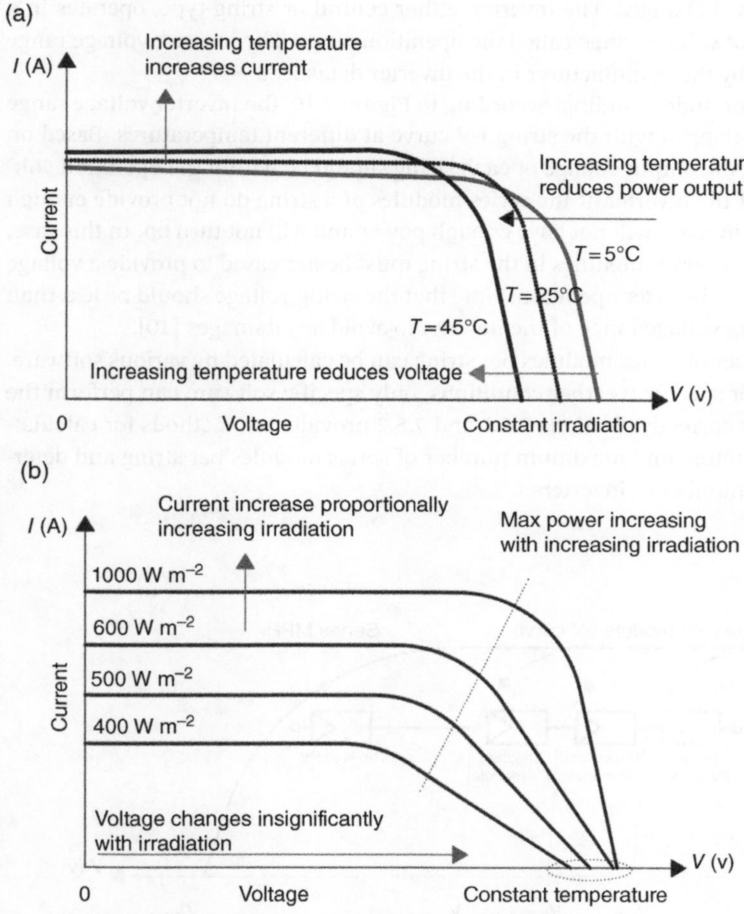
(a)
Increasing temperature
increases current
I (A)
Increasing temperature
reduces power output
Current
T = 5°C
T = 25°C
T = 45°C
Increasing temperature reduces voltage
V (v)
0
Voltage
Constant irradiation
(b)
Current increase proportionally
increasing irradiation
I (A)
Max power increasing
with increasing irradiation
1000 W m-2
600 W m-2
500 W m-2
400 W m-2
Current
Voltage changes insignificantly
with irradiation
V (v)
0
Voltage
Constant temperature

Figure 7.8 The effect of irradiation intensity and temperature on the I-V curve [8]. (a) The I-V characteristics at different temperatures, and (b) The I-V characteristics at different irradiation intensities and a constant temperature. *Source:* Based on Amiry et al. [8].

b) By reducing the irradiation level of a PV module, the current and output power of the module is reduced.

By connecting several PV modules in series, a string is created and the I-V curve becomes similar to the one shown in Figure 7.9. The series connection of PV modules leads to an increase in the string voltage. The I-V curve of a PV module is a function of environmental parameters such as temperature, irradiation, and load connected to the module. Therefore, the I-V curve of each string is a function of these parameters too.

An inverter uses internal algorithms to track the MPP of the module output when the load changes. The inverter, either central or string type, operates in a specific input voltage range called the operation range. The inverter voltage range is specified by the manufacturer in the inverter datasheet.

For a better understanding, according to Figure 7.10, the inverter voltage range curve is overlapped with the string I-V curve at different temperatures. Based on Figure 7.10, the output voltage of each string should be within the operation voltage range of the inverter. If the series modules of a string do not provide enough voltage, the inverter will not have enough power and will not turn on. In this case, the number of series modules in the string must be increased to provide a voltage in the range of inverter operation. Note that the string voltage should be less than the operating voltage range of the inverter to avoid any damages [10].

The number of series modules per string can be calculated by various software. However, for specific weather conditions, only specific software can perform the calculations correctly. Sections 7.5.1 and 7.5.2 provide two methods for calculating the minimum and maximum number of series modules per string and determining the number of inverters.

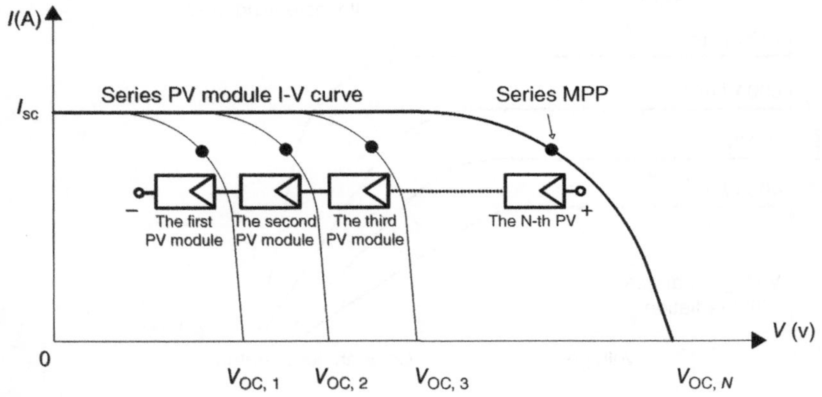
I(A)
Series PV module I-V curve
Series MPP
I_{sc}
−
The first PV module
The second PV module
The third PV module
The N-th PV
+
V (v)
0
$V_{OC,1}$
$V_{OC,2}$
$V_{OC,3}$
$V_{OC,N}$

Figure 7.9 The I-V curve of PV modules connected in series to create a string. *Source:* Adapted from Rakesh et al. [9] and Deutsche Gesellschaft für Sonnenenergie (DGS) [10].

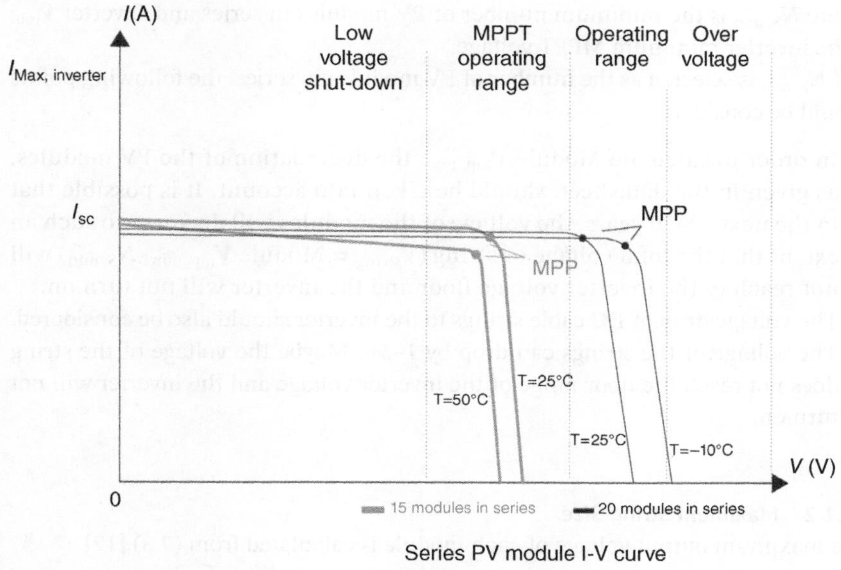
I(A)
$I_{Max, inverter}$
I_{sc}
Low voltage shut-down
MPPT operating range
Operating range
Over voltage
MPP
MPP
T=25°C
T=50°C
T=25°C
T=−10°C
V (V)
0
15 modules in series
20 modules in series
Series PV module I-V curve

Figure 7.10 The inverter voltage range with the string I-V curve at different temperatures. *Source:* Modified from Dupré [11].

7.5.1 Method 1

7.5.1.1 Minimum String Size

The minimum output voltage of each module (at the installation site) is calculated from (7.1) [12].

$$\text{Module}\, V_{\text{mp,min}} = V_{\text{mp}(70°\text{C})} = V_{\text{mp}} \times \left(1 + \frac{T_{\max} \cdot \Delta T}{100\%}\right) \tag{7.1}$$

In (7.1), Module $V_{\text{mp, min}}$ is the minimum module voltage expected at the highest site temperature, V_{mp} is the rated module voltage at max power, $T_{\max}$ is temperature coefficient at maximum expected temperature, ΔT is temperature variance between standard test condition (STC) and the expected maximum temperature [12]. Normally, V_{mp} is determined at 70 °C. However, for calculating V_{mp} in temperatures other than 70 °C, the maximum annual ambient temperature should be added with 25 °C.

The minimum number of modules in a string is obtained from (7.2) [13].

$$N_{\text{S,min}} \geq \frac{\text{Inverter}\, V_{\min}}{\text{Module}\, V_{\text{mp,min}}} \tag{7.2}$$

where $N_{S, min}$ is the minimum number of PV modules in series and Inverter V_{min} is the inverter minimum MPPT voltage.

If $N_{S, min}$ is selected as the number of PV modules in series, the following points should be considered:

1) In order to calculate Module $V_{mp, min}$, the degradation of the PV modules, as given in the datasheet, should be taken into account. It is possible that in the next 10–20 years, the voltage of the modules will decrease to such an extent that the total voltage of string (V_{string} = Module $V_{mp, min} \cdot N_{S, min}$) will not reach to the inverter voltage floor and the inverter will not turn on.
2) The voltage drop of DC cable strings to the inverter should also be considered. The voltage of the strings can drop by 1–3%. Maybe the voltage of the string does not reach the floor range of the inverter voltage and the inverter will not turn on.

7.5.1.2 Maximum String Size

The maximum output voltage of each module is calculated from (7.3) [12].

$$\text{Module}\, V_{OC,max} = V_{OC(-10°C)} = V_{OC}\left(1 + \frac{T_{min} \cdot \Delta T}{100\%}\right) \tag{7.3}$$

where Module $V_{OC, max}$ is the maximum module voltage corrected for the lowest ambient temperature, V_{OC} is the rated open-circuit voltage of the PV module, T_{min} is temperature coefficient at minimum temperature, and ΔT is temperature variance between STC and minimum temperature. Normally, V_{OC} is determined at −10 °C. However, for the regions such as Alpine zone, the minimum temperature subtracted by 25 °C is used to calculate V_{OC}.

The maximum number of modules in a string is obtained from (7.4) [13].

$$N_{S,max} \leq \frac{\text{Inverter}\, V_{max}}{\text{Module}\, V_{OC-max}} \tag{7.4}$$

where $N_{S, max}$ is the maximum number of PV modules in series and Inverter V_{max} is the inverter maximum allowable voltage.

The optimum number of PV modules must not be less than the minimum number of PV modules per string and must not exceed the maximum number. As a rule of thumb, the more PV modules per string there are, the more viable is the planning of the PV array.

7.5.1.3 Determining Maximum Current of a PV Module

The PV module current is the highest at high temperatures. The maximum PV module current is obtained from (7.5) [12],

$$\text{Module}\, I_{\text{DC,max}} = I_{\text{DC}(70°\text{C})} = I_{\text{sc}}\left(1+\frac{T_{\max}\cdot\Delta T}{100\%}\right) \tag{7.5}$$

In (7.5), Module $I_{\text{DC, max}}$ is maximum string current, I_{DC} is the short-circuit current of the PV module, $T_{\max}$ is temperature coefficient at maximum expected temperature, and ΔT is temperature variance between STC and maximum expected temperature.

7.5.1.4 Determining Number of Inverters

Inverters take a considerable portion of LS-PVPP initial cost. Therefore, determining the inverter size, in terms of the balance between energy efficiency and inverter investment, is important for the investor. Based on conventional practice [14], the number of inverters is obtained such that the nominal input power of the inverter is approximately equal to the nominal output power of PV array, i.e. $P_{\text{inv, nom}} = P_{\text{PV, nom}}$.

The distribution of the annual solar irradiation, environmental conditions, type of mounted raking PV, and PV losses are the main factors that affect the number of inverters. The number and power rating of inverters are determined by the overall power of the PV plant. A reliable method is to calculate the DC power via the inverter's nominal efficiency from the AC nominal power. The ratio of rated inverter power to the PV array power rating is known as the inverter sizing factor or nominal power ratio (NPR) [14].

$$\text{NPR} = \frac{P_{\text{inv,nom}}}{P_{\text{PV,nom}}} = \frac{G_{\text{Th}}}{G_{\text{STC}}} \tag{7.6}$$

where $P_{\text{PV, nom}}$ is the rated PV installed power, G_{Th} is the nominal irradiance level, and $G_{\text{STC}} = 1000\,\text{W/m^2}$. NPR is the inverter downsize coefficient. When $0<\text{NPR}<1$, the inverter is undersized. For $\text{NPR}>1$ the inverter is being oversized and will have higher power rating that the PV installed power [14].

The number of inverters (N_{i}) can be determined as follows:

$$N_{\text{i}} = \frac{P_{\text{inv,nom-total}}}{P_{\text{inv,nom-expected}}} \tag{7.7}$$

where $P_{\text{inv, nom-total}}$ is the total power of the inverter that is required for the entire power plant and $P_{\text{inv, nom-expected}}$ is the expected power of the inverter.

For example, for an LSPVP with 100 MW capacity, the NPR and $P_{\text{inv, nom}}$ can be determined using (7.6) as follows.

a) Irradiance = $1000\,\text{W}\,\text{m}^{-2}$ (STC[1]) → NPR = 1 → $P_{\text{inv, nom}-\text{total}} = 100$ MW

   $P_{\text{inv, nom}-\text{expected}} = 150$ kW → $N_{\text{i}} \cong 666$ unit

b) Irradiance = $600\,\text{W}\,\text{m}^{-2}$ → NPR = 0.6 → $P_{\text{inv, nom}-\text{total}} = 60$ MW

   $P_{\text{inv, nom}-\text{expected}} = 150\,\text{kW}$ → $N_i \cong 400$ unit

The power $P_{\text{inv, nom}}$ specifies the DC power that must be injected to the inverter input so that the desired AC power can be supplied to the grid. $P_{\text{inv, nom}}$ is calculated from (7.8) [13].

$$P_{\text{inv,nom}} = \frac{S_{\text{AC}}.\cos\varphi}{\eta} \tag{7.8}$$

where S_{AC} is AC active power and shows how much power, at power factor cos φ, should be injected to the grid, and η is the inverter efficiency. Care must also be taken in the calculation of the inverter efficiency that is affected by the PV array voltage [13].

Method 1 is an experimental method to determine the inverter size. In [14], a method for determining the optimum grid-connected PV inverter size is presented.

7.5.2 Method 2

In this method, using the algorithm of Figure 7.11, the following information is calculated [15]:

1) Maximum number of PV modules in series per inverter ($N_{\text{S, max}}$)
2) Maximum number of PV modules connected in parallel ($N_{\text{P, max}}$)
3) Number of inverters (N_{i})
4) Number of PV modules ($N_{\text{PV, final}}$)
5) Total installed capacity of the PV power plant ($P_{\text{installed}}$)

Note that in this method, by choosing the maximum number of PV modules connected in series, the number of required inverters is reduced. However, in terms of efficiency, this is not always the best option since an inverter is usually more efficient when it is operating around its rated power [15].

7.6 Size of PV Plant DC Side

Many parameters are involved in determining the size of DC side of a PV power plant. The main parameters are as follows:

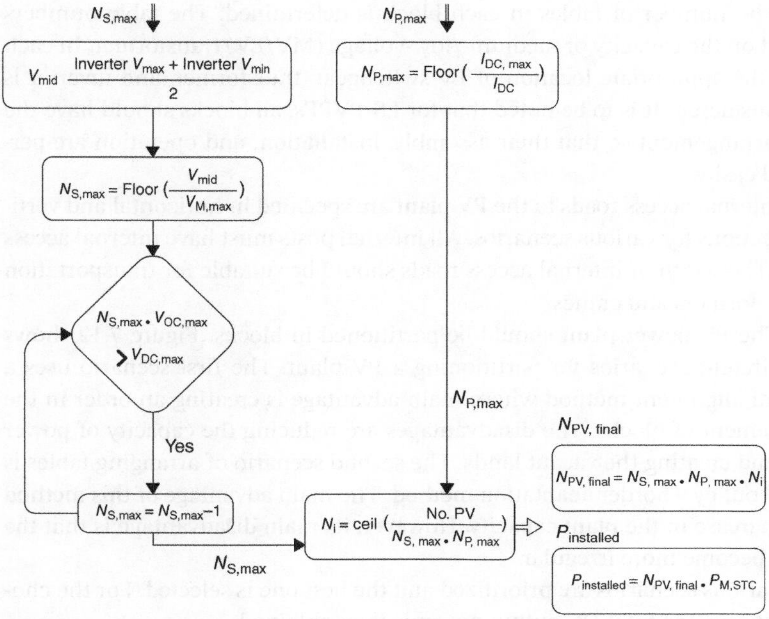

$N_{S,max}$
$N_{P,max}$
$V_{mid} = \frac{\text{Inverter } V_{max} + \text{Inverter } V_{min}}{2}$
$N_{P,max} = \text{Floor}\left(\frac{I_{DC,max}}{I_{DC}}\right)$
$N_{S,max} = \text{Floor}\left(\frac{V_{mid}}{V_{M,max}}\right)$
$N_{S,max} \cdot V_{OC,max} > V_{DC,max}$
$N_{P,max}$
$N_{PV,final}$
Yes
$N_{PV,final} = N_{S,max} \cdot N_{P,max} \cdot N_i$
$N_{S,max} = N_{S,max} - 1$
$N_i = \text{ceil}\left(\frac{\text{No. PV}}{N_{S,max} \cdot N_{P,max}}\right)$
$P_{installed}$
$N_{S,max}$
$P_{installed} = N_{PV,final} \cdot P_{M,STC}$

Figure 7.11 Maximum number of PV modules in series algorithm [15]. ($P_{m,STC}$ is the nominal power PV module at STC.) *Source:* Based on Roca Rubí [15].

- Allowed power capacity of the network to which the PV plant is connected
- Useful land area and the land geometry
- Technology, output power, and efficiency of PV modules
- Optimal angle and spacing at the site
- Type of structure, i.e. fixed or tracked type
- Financial budget of the investor

To determine the size of a PV power plant, seven steps are performed as explained next.

Step 1: Technical specifications of the PV modules, optimal angle, shading distance, table direction (south or east-west), the number of modules in each string, and the coordinates of the corners of the site are determined.

Step 2: The number of PV modules is specified according to the maximum available financial budget.

Step 3: The table dimensions are calculated. The size of each table depends on the number of modules per string. Considering a proper table size facilitates defects identification and maintenance of PV modules during the plant operation.

Step 4: The number of tables in each block is determined. The table numbers depend on the capacity of medium-/low-voltage (MV/LV) transformer. In each block, the appropriate location of LV switchgear, transformer, and inverter is also considered. It is to be noted that for LS-PVPPs, all blocks should have the same arrangement so that their assembly, installation, and operation are performed easily.

Step 5: Internal access roads to the PV plant are specified in horizontal and vertical directions for various scenarios. All internal posts must have internal access roads. The width of internal access roads should be suitable for transportation of transformers and cranes.

Step 6: The PV power plant should be partitioned in blocks. Figure 7.12 shows two different scenarios for partitioning a PV plant. The first scenario uses a uniform alignment method whose main advantage is creating an order in the arrangement of blocks. The disadvantages are reducing the capacity of power plant and creating the vacant lands. The second scenario of arranging tables is carried out by a border adaptation method. The main advantage of this method is an increase in the plant capacity. However, its main disadvantage is that the blocks become more irregular.

Step 7: Various scenarios are prioritized and the best one is selected. For the chosen scenario, the total DC output power is then obtained.

7.7 DC Cables

7.7.1 Criteria

PV DC cables are single-conductor wires that connect PV modules to an inverter. The DC cables are exposed to indirect sunlight and ambient temperature during the day. Therefore, the insulation of DC cables must be resistant to weather

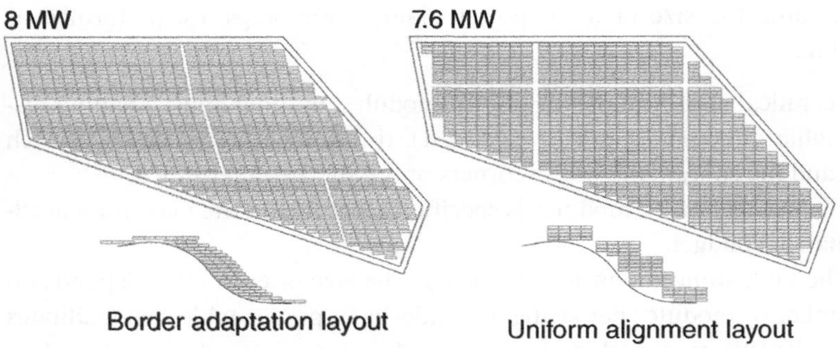
8 MW
7.6 MW
Border adaptation layout
Uniform alignment layout

Figure 7.12 Two different scenarios for partitioning a PV plant.

conditions, ultraviolet (UV) rays, abrasion, brine, acids, and alkaline solutions. The DC cables must last between 25 and 30 years in harsh environmental conditions. The wires should be chosen in two colors for easy service, e.g. red for positive and blue for negative. Technical specifications and methodology of selecting DC cables are shown in Figure 7.13.

7.7.2 DC Cables Cross Section

To determine the size of DC cables, the following parameters must be specified:

- Installation conditions including environment temperature and number of adjacent cables
- Cable current capacity
- Cable voltage range
- Percentage of voltage drop
- Percentage of power loss
- Cable short-circuit current

In Sections 7.7.2.1–7.7.2.4, it is discussed how to determine the above parameters.

7.7.2.1 Current Capacity

In the string inverter configuration, a number of strings are connected in parallel with the inverter, while in the central inverter configuration, the strings are connected in parallel to a combiner box. For the latter, DC cables with a larger cross section are connected between a number of combiner boxes and the central inverter. For both configurations, to determine the DC cable cross section, the following must be taken into consideration:

- Cable current should not be less than the designed current value; and
- Cable voltage drop should be less than a certain value.

After determining the optimal number of modules in each string, the cable current flow is obtained as follows.

$$I_n = I_{sc} \cdot N_{string} \cdot S_F \tag{7.9}$$

where I_{sc} is the short-circuit current of PV module under STC, and S_F is the safety coefficient and is generally considered between 1.2 and 1.3, depending on the installation requirements. To calculate the current of each cable connected to the combiner box, N_{string} is set to 1. To find the current of combiner box output cable, N_{string} is set to the number of parallel strings connected the combiner box.

The datasheet of a DC cable provides its current capacity, resistance, cross section, and conductor materials. The technical specifications of a DC cable are given

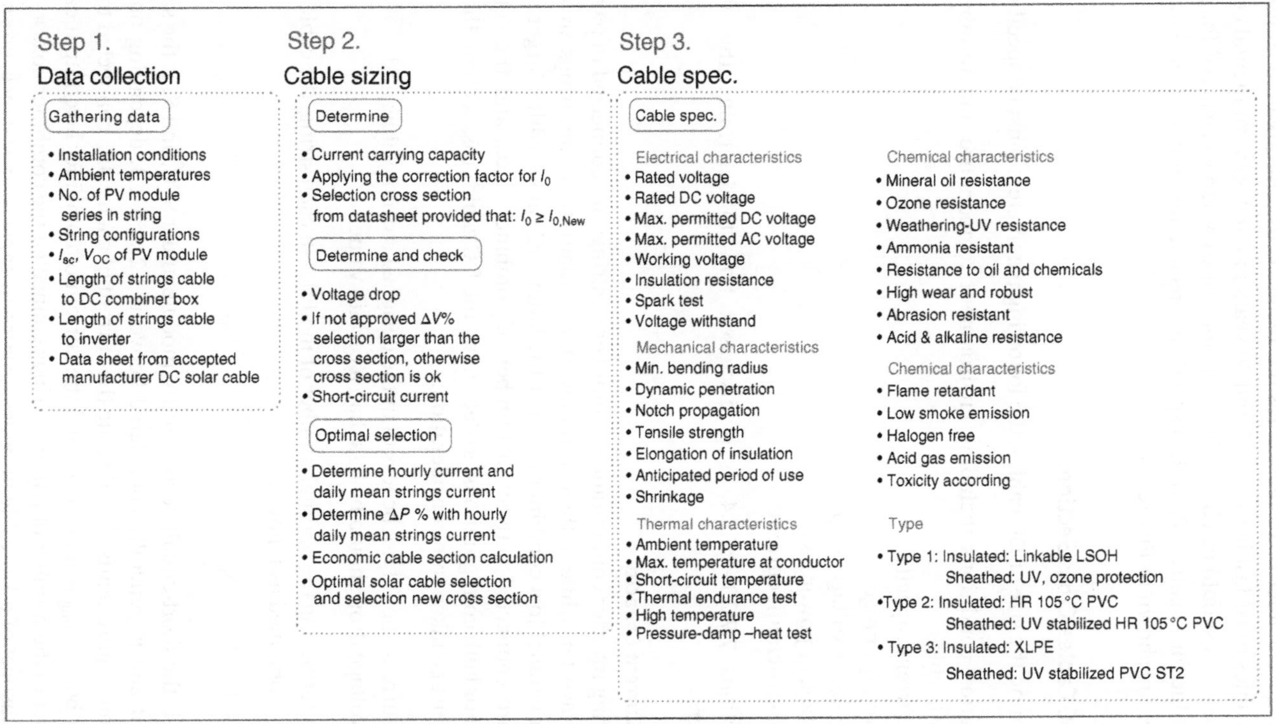
Step 1.
Data collection
Gathering data
• Installation conditions
• Ambient temperatures
• No. of PV module series in string
• String configurations
• I_{sc}, V_{OC} of PV module
• Length of strings cable to DC combiner box
• Length of strings cable to inverter
• Data sheet from accepted manufacturer DC solar cable
Step 2.
Cable sizing
Determine
• Current carrying capacity
• Applying the correction factor for I_0
• Selection cross section from datasheet provided that: $I_0 \geq I_{0,New}$
Determine and check
• Voltage drop
• If not approved ΔV% selection larger than the cross section, otherwise cross section is ok
• Short-circuit current
Optimal selection
• Determine hourly current and daily mean strings current
• Determine ΔP % with hourly daily mean strings current
• Economic cable section calculation
• Optimal solar cable selection and selection new cross section
Step 3.
Cable spec.
Cable spec.
Electrical characteristics
• Rated voltage
• Rated DC voltage
• Max. permitted DC voltage
• Max. permitted AC voltage
• Working voltage
• Insulation resistance
• Spark test
• Voltage withstand
Mechanical characteristics
• Min. bending radius
• Dynamic penetration
• Notch propagation
• Tensile strength
• Elongation of insulation
• Anticipated period of use
• Shrinkage
Thermal characteristics
• Ambient temperature
• Max. temperature at conductor
• Short-circuit temperature
• Thermal endurance test
• High temperature
• Pressure-damp –heat test
Chemical characteristics
• Mineral oil resistance
• Ozone resistance
• Weathering-UV resistance
• Ammonia resistant
• Resistance to oil and chemicals
• High wear and robust
• Abrasion resistant
• Acid & alkaline resistance
Chemical characteristics
• Flame retardant
• Low smoke emission
• Halogen free
• Acid gas emission
• Toxicity according
Type
• Type 1: Insulated: Linkable LSOH
Sheathed: UV, ozone protection
•Type 2: Insulated: HR 105 °C PVC
Sheathed: UV stabilized HR 105 °C PVC
• Type 3: Insulated: XLPE
Sheathed: UV stabilized PVC ST2

Figure 7.13 Technical specifications and methodology of DC cables selection. *Source:* Adapted from Berwal et al. [16] and Garcia et al. [17].

in Table 7.3. Each cross section has a specific current carrying capacity (I_o) that is usually given at 30 °C in the open air.

Depending on the temperature conditions and the cable installation in ground or air, the actual amount of I_o is calculated as

$$I_{o,\text{NEW}} = \frac{I_o}{K} \tag{7.10}$$

where I_o is the cable current carrying capacity from datasheet, $I_{o,\text{NEW}}$ is the current carrying capacity corresponding to the installation conditions, and K is the correction factor obtained from [19]:

$$K = T_\text{F}.G_\text{F} \tag{7.11}$$

In (7.11), T_F is the correction factor used for ambient temperature, and G_F is the reduction factor used for more than one circuit. For a copper DC cable that passes through a conduit or trunking system and is exposed to a temperature of 70°, the correction factor is:

$$K = 0.58 \cdot 0.9 = 0.52 \tag{7.12}$$

The cable insulation, material and installation method, temperature conditions, and the number of adjacent cables affect the temperature correction factor. The International Electrotechnical Commission (IEC) 60364-5-52 standard accurately determines the correction factor.

7.7.2.2 Voltage Drop

After determining $I_{o,\text{NEW}}$, depending on the type of cable, an appropriate cross section is selected. At this step, the cable cross section cannot be determined before computing the percentage of voltage drop for the cable length.

The acceptable voltage drop depends on the requirements and regulations in different countries. Based on the IEC standard 60364-5-52, the acceptable voltage drop is between 1 and 3%. Higher voltage drop leads to higher system losses [20]. If the percentage of voltage drop is inappropriate, the cross section of cable is increased to an acceptable value. The percentage of cable voltage drop is calculated as follows [21].

$$\Delta V\% = \frac{2ILR}{V_n} \times 100 \tag{7.13}$$

where L is the cable length, I is the nominal current, and R is the cable resistance per meter.

The nominal voltage of each string is calculated from

$$V_n = V_\text{MPP} \cdot N_\text{module} \tag{7.14}$$

Table 7.3 The technical specifications of a DC solar cable [18].

Cross section (mm^2)	American wire gauge (AWG)	Conductor diameter max. (mm)	Weight ($kg\,km^{-1}$)	Permissible tensile force max. (N)	Conductor resistance at 20°C max. ($\Omega\,km^{-1}$)	Current carrying capacity for single cable free in air (60 °C ambient temp.) (A)	Current carrying capacity for single cable on a surface (60 °C ambient temp.) (A)	Short-circuit current (1 s from 90 to 250 °C) (kA)
1×2.5	14	1.9	46	38	8.21	41	39	0.36
1×4	12	2.4	61	60	5.09	55	52	0.57
1×6	10	2.9	80	90	3.39	70	67	0.86
1×10	8	4	122	150	1.95	98	93	1.43
1×16	6	5.6	200	240	1.24	132	125	2.29
1×25	4	6.4	290	375	0.795	176	167	3.58
1×35	2	7.5	400	525	0.565	218	207	5.01
1×50	1	9	560	750	0.393	276	262	7.15
1×70	2/0	10.8	750	1050	0.277	347	330	10.01
1×95	3/0	12.6	970	1425	0.21	416	395	13.59
1×120	4/0	14.2	1220	1800	0.164	488	464	17.16
1×150	300MCM	15.8	1500	2250	0.132	566	538	21.45
1×185	350MCM	17.4	1840	2775	0.108	644	612	26.46
1×240	500MCM	20.4	2400	3600	0.082	775	736	34.32

Source: Based on Saldaña Pizarraya [18].

where N_{module} is the number of modules connected in series to a string, and V_{MPP} is the voltage of one module at the MPP.

The cable resistance is obtained from the datasheet for the given cross section and cable material. The length of positive and negative wires is multiplied by 2 to obtain the total cable length. If the percentage of voltage drop is not appropriate, the cross section of the cable is increased to bring the voltage drop to an acceptable range.

7.7.2.3 Power Loss

To select a proper cross section, the percentage of DC cable power loss is calculated from (7.15) [20, 21].

$$\Delta P\% = \frac{2RI^2L}{P_n} \times 100 \tag{7.15}$$

where P_n is the total $P_{m,\,STC}$ (nominal power PV module at STC) of modules connected in series (w), L is one-way length of DC cable (m), R is resistance of DC solar cable (ohm per m), and I is maximum continuous hourly current (A). Various cable cross sections are compared in terms of their purchase price, performance, and revenue resulted from reducing the power losses. Then, an optimal cable cross section is determined.

The current flowing in a cable depends on the sun radiation and other environmental conditions. The amount of current for each string is calculated for a period of one year by a simulation software, and the total power loss percentage per year is calculated from (7.15) [20, 21].

7.7.2.4 Short-circuit Current

To choose the type of cable, short-circuit current capability is a determining factor to be considered. In the event of short circuit, the current suddenly increases for several cycles and then decreases until the protection system operates. The short-circuit current of a DC cable is obtained from [21]:

$$I_{sc} = \frac{K \times S}{\sqrt{t}} \tag{7.16}$$

where S is the cable cross section, t is short-circuit duration that is 1 second for MV, and K is a constant coefficient. The value of K for polyvinyl chloride (PVC) and cross-linked polyethylene (XLPE) coated copper and aluminum conductors varies based on their initial and maximum operating temperatures. The IEC 6034-5-54 standard provides the details on how to calculate K. According to (7.17), the DC short-circuit current must be greater than the string short-circuit current.

$$I_{sc} > I_{sc-string} \tag{7.17}$$

In (7.17), $I_{sc-string}$ is the short-circuit current of PV module given in the solar panel datasheet.

7.8 DC Combiner Box

It is important to determine the number of combiner boxes and their components. In LS-PVPPs, the main function of DC box is to combine the strings and arrays to reduce the cabling time and expense. Moreover, the combiner box protects the DC section against faults, over-voltages, and over-currents, which lead to increased internal network reliability.

Some inverters are equipped with the DC fuse, surge arrester, and switches. In these inverters, the box combiner could be eliminated. In such cases, only the DC equipment inside the inverter is examined.

To determine the size and number of DC combiner boxes, the following should be determined:

- Inverter topologies and configuration
- Inverter rated capacity
- Number of MPPT inputs per inverter
- Combiner box distance
- Percentage of voltage drop
- Site geometry and weather conditions
- Maximum open-circuit voltage of each string at the lowest temperature
- Total maximum short-circuit current of the strings at maximum radiation
- Cross section of input and output cables, and number of inputs and outputs
- Features of monitoring system
- Internal components of the combiner box

The location map of the internal components of the DC combiner box is shown in Figure 7.14. For each string, the input terminals (positive and negative), fuse holder, fuse base, string diode, surge arrester, and DC switch are shown. In some designs, instead of a DC switch, an automatic switch is used.

The combiner box should be easily accessible so that the operating technician can easily inspect it and replace any equipment in case of failures. The combiner boxes are generally installed outdoors, and thus they should be UV resistant. Moreover, the combiner box should be noncombustible and waterproof. They should generally have protection class IP54 and preferably IP65. In the following, the main internal equipment of the combiner box is discussed [22].

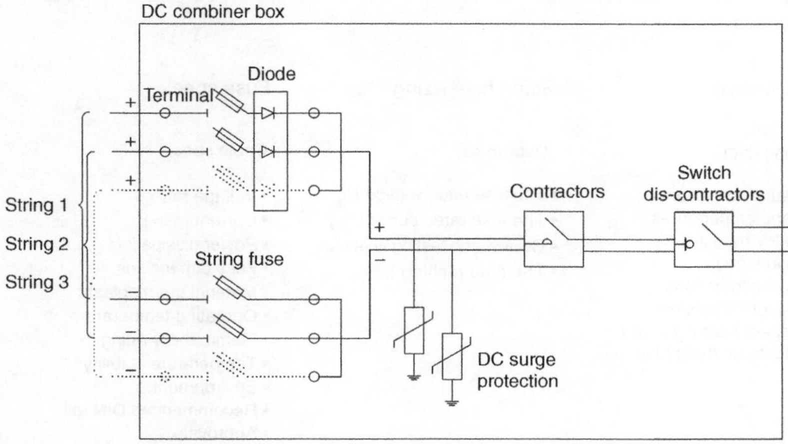
DC combiner box
Diode
Terminal
String 1
String 2
String 3
String fuse
Contractors
Switch
dis-contractors
DC surge
protection

Figure 7.14 DC combiner box.

7.9 String Diode

The string diode prevents reverse current to flow to the PV module. Reverse current can occur in the event of malfunctions such as shading, breakdowns, and faults. The diode must be able to carry a current of at least 1.25 times of the maximum short-circuit current of the string. It is always installed on the non-ground side. The rated voltage of the string diode must be higher than the maximum open-circuit voltage of the string, i.e. [23]:

$$V_{ID} \geq 2 \cdot V_{OC,STC} \tag{7.18}$$

In (7.18), V_{ID} is the reverse voltage of string diode and $V_{OC,\,STC}$ is the maximum operating open-circuit voltage of the string.

Depending on the local regulations, complete protection against reverse current is not required in some countries. In such cases, the string diode for reverse current protection is removed from the combiner box.

7.10 Fuse

For safe, reliable, and long-lasting operation, the DC side of a PV power plant requires a fuse with an appropriate current rating. Fuses are required to protect cables and PV modules from line-to-line, line-to-ground, and mismatch faults. In cases where one side of the string is grounded, the fuse must be installed on the

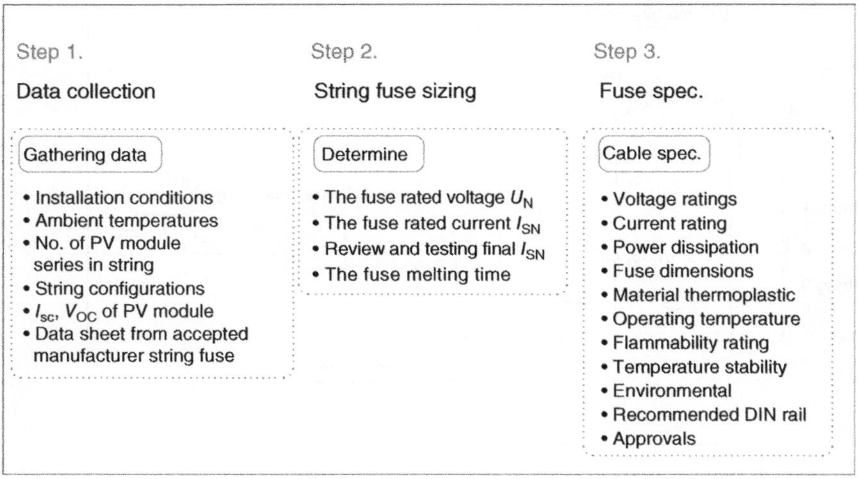
Step 1.
Data collection
Gathering data
• Installation conditions
• Ambient temperatures
• No. of PV module series in string
• String configurations
• I_{sc}, V_{OC} of PV module
• Data sheet from accepted manufacturer string fuse
Step 2.
String fuse sizing
Determine
• The fuse rated voltage U_N
• The fuse rated current I_{SN}
• Review and testing final I_{SN}
• The fuse melting time
Step 3.
Fuse spec.
Cable spec.
• Voltage ratings
• Current rating
• Power dissipation
• Fuse dimensions
• Material thermoplastic
• Operating temperature
• Flammability rating
• Temperature stability
• Environmental
• Recommended DIN rail
• Approvals

Figure 7.15 The procedure of fuse selection. *Source:* Based on Haas [24].

other side. For each PV string, a DC fuse is considered for positive and negative cables.

To select an appropriate fuse for a string, information about the PV module from its datasheet and the modules installation conditions are required. In some countries the aR class fuses are used for PV systems. The aR class fuses are not suitable for PV systems since they are prone to fire. Figure 7.15 shows the procedure for selecting a fuse and provides the required technical specifications. The fuse selection steps are also detailed in the following sections [24].

7.10.1 Rated Voltage

The voltage of a string depends on the number of string modules and the ambient temperature. The rated fuse voltage U_n must satisfy [24]:

$$U_n > V_{\text{OC,STC}} \cdot N_{\text{string}} \cdot 1.2 \tag{7.19}$$

In (7.19), $V_{\text{OC, STC}}$ is the open-circuit voltage of PV modules and N_{string} is the number of PV modules in a string.

7.10.2 Rated Current

The rated current of a fuse at 25–30 °C is provided in the datasheet. To calculate the actual rated current of a fuse, a correction coefficient for the ambient temperature, the number of adjacent fuses, and the load variations are considered. The nominal current of a fuse I_{SN} is obtained from [24].

$$I_{SN} = \frac{I_{sc,STC}}{k_T \cdot k_L \cdot k_H} \tag{7.20}$$

where $I_{sc,\,STC}$ is the short-circuit current, k_T is the ambient temperature coefficient from datasheet, e.g. at 60 °C $k_T = 0.84$, k_L is the alternating load factor that is normally considered as 0.9, and k_H is the derating factor for high number of adjacent fuses, e.g. for a group of three fuses, it is equal to 1. I_{SN} can also be obtained from:

$$I_{SN} = I_{sc,STC} \cdot k_{SN} \tag{7.21}$$

where $1.4 < k_{SN} < 2$, and it is generally considered at least 1.5 for desert areas and 1.6 for cold mountainous areas.

7.10.3 Fuse Testing

Fuse manufacturers consider a larger current than I_{SN} (I_{FD}) in the datasheet. The testing current of a fuse is obtained as follows.

$$I_{testing} = I_{FD} \cdot k_T \cdot k_L \cdot k_H \tag{7.22}$$

$$I_{DC,max} = (I_{sc} \cdot (1 + \left(\Delta\vartheta \cdot \text{temp.coeff.of } I_{sc,string}\right) \cdot k_{Ir} \tag{7.23}$$

where I_{FD} is the fuse rated current given in the datasheet, $I_{DC,\,max}$ is the I_{sc} at maximum ambient temperature, and k_{Ir} is the maximum irradiance, e.g. it is 1.2 for 1200 W/m^2. The fuse testing current should satisfy the following condition.

$$I_{testing} > I_{DC,max} \tag{7.24}$$

7.10.4 Melting Time

In the final step of fuse selection, the fuse operating time following the short circuit is evaluated. The short-circuit current and the maximum short-circuit current of the string are plotted in the fuse time-current characteristic, where their intersection is the fuse operating time. The string short-circuit current is calculated as follows [24]:

$$\text{String short circuit residual current } I_{sc,string} = I_{sc,MOD} \cdot \left(N - 1\right) \tag{7.25}$$

where $I_{sc,\,MOD}$ is the short-circuit current of the PV module and N is the number of residual parallel strings.

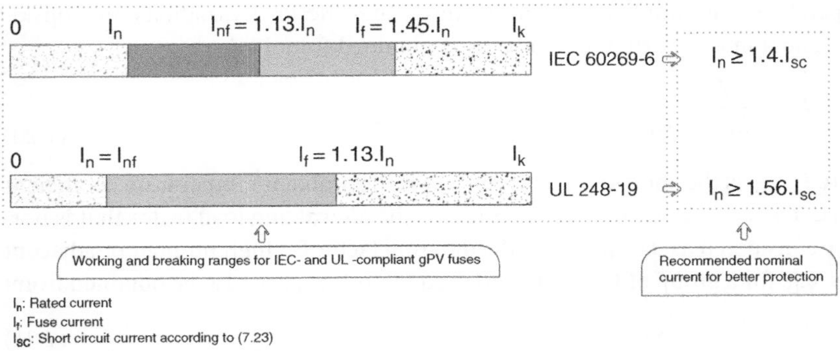
0
In
Inf = 1.13.In
If = 1.45.In
Ik
IEC 60269-6
In ≥ 1.4.Isc
0
In = Inf
If = 1.13.In
Ik
UL 248-19
In ≥ 1.56.Isc
Working and breaking ranges for IEC- and UL -compliant gPV fuses
Recommended nominal current for better protection
In: Rated current
If: Fuse current
Isc: Short circuit current according to (7.23)

Figure 7.16 Minimum rated current for string fuses according to the IEC and UL standards. *Source:* Based on Haas [24].

Finally, as shown in Figure 7.16, the IEC 60269-6 and UL 248-19 standards recommend that for reliable protection, the rated current should be considered for the fuse selection [23].

7.11 Surge Arrester

The land of a LS-PVPP is usually large. Therefore, there is a high possibility of lightning strikes to the plant site during its operation. Therefore, protection of PV plants against lightning strikes is required to prevent catastrophic damages to the electronic equipment. Lightning strikes may also endanger the safety of personnel. A person at a distance less than 60 feet from the lightning point can be killed by the indirect lightning wave.

When a lightning strikes a PV plant, it creates transient currents and voltages in the power network. Circulation of transient currents and voltages in the equipment can cause insulation failures in the electrical and electronic equipment. The equipment that may fail due to the transients resulting from lighting include inverters, switches, monitoring systems, building installations, and other expensive equipment.

To prevent high energy of lighting to pass through the power network of a PV plant, voltage surges must have a path to the ground. A surge protective device (SPD) is used for each group of strings connected to a DC box to protect the equipment against voltage surges. The surge protection device discharges extra currents to the ground and prevents the entry of destructive voltages to electrical and electronic equipment.

7.12 DC Switch

The DC switch is an equipment that manually separates the entire PV array from inverter. The array separation may be required for safety reasons or maintenance. The DC switch is located close to the inverter or in the combiner box. When choosing a DC switch, the following parameters are taken into consideration [25]:

- Rated insulation voltage U_i,
- Rated operating voltage U_e, and
- Rated operating current I_e.

The rated insulation voltage U_i should never be less than the open-circuit voltage V_{oc}. The rated operating voltage U_e of the switch should always be greater than the voltage level at which the current failure occurs. The value of U_e depends on the dielectric strength and the distance between the internal conductor and the creep. The rated operating current I_e of the switch should be greater than the total short-circuit currents of the PV array. If any of the parameters U_i, U_e, or I_e is not selected properly, it can cause switch malfunction that in turn poses safety hazards to the end user.

To calculate I_e, the switch derating factor is first determined as [25]:

$$\text{Derating factor} = \sqrt{\frac{T_{\text{max normal}} + \Delta_{\text{max}} T - T_{\text{amb}}}{\Delta_{\text{max}} T}} \tag{7.26}$$

where $T_{\text{maxnormal}}$ is the maximum permitted average temperature under normal conditions (35° C), $\Delta_{\text{max}} T$ is the maximum allowable temperature rise that is 70 ° C, and T_{amb} is the ambient temperature.

Figure 7.17 shows a derating factor calculated for different ambient temperatures. The switch factors depend on the technical specifications provided by the manufacturers. From the curve shown in Figure 7.17 provided by the manufacturer, the minimum current of the switch is obtained from

$$I_e > \frac{I_{\text{SCMT}}\ \text{Array PV}}{K_T} \tag{7.27}$$

In (7.27), I_{SCMT} Array PV is the total short-circuit current of the array. The IEC 60947-3 standard provides the technical specifications of various types of switches. For example, the DC-21B switch is suitable for average overloads where the shutoff operation is less likely to occur. This type of switch is usually employed for PV power plants. For the switch selection, there are noticeable differences between the recommendations provided by the UL 98 and UL 508 standards and the IEC 60947 and National Electrical Code (NEC) standards [25].

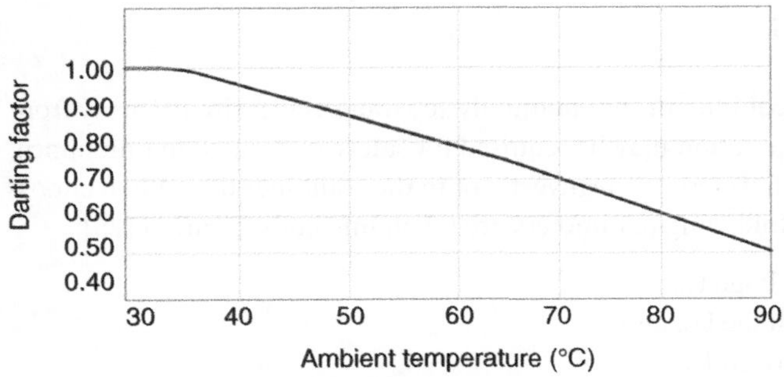
Darting factor
1.00
0.90
0.80
0.70
0.60
0.50
0.40
30
40
50
60
70
80
90
Ambient temperature (°C)

Figure 7.17 Derating factor for different ambient temperatures. *Source:* Based on ABB Oy low voltage products [25].

7.13 Conclusion

In this chapter, different components of the DC side of a PV power plant including their design methodology were presented and discussed. The presented components may be subject to changes depending on the type of implementation or contract and the terms of the project. In this chapter, we have also presented a simple, concise, and useful approach for selecting and calculating the DC side components of a PV power plant.

Note

1 STC: Standard test condition of PV modules: irradiance = 1000 W m^{-2}, cell temperature = 25 °C, and air mass = 1.5.

References

1 Falvo, M.C. and Capparella, S. (2015). Safety issues in PV systems: design choices for a secure fault detection and for preventing fire risk. *Case Studies in Fire Safety* 3: 1–16.

2 Udayakumar, M.D., Anushree, G., Sathyaraj, J., and Manjunathan, A. (2021). The impact of advanced technological developments on solar PV value chain. *Materials Today: Proceedings* 45: 2053–2058.

3 El-Bayeh, C.Z., Alzaareer, K., Brahmi, B. et al. (2021). An original multi-criteria decision-making algorithm for solar panels selection in buildings. *Energy* 217: 119396.

4 Cabrera Tobar, A.K. (2018). Large scale photovoltaic power plants: configuration, integration and control. Universitat Politecnica de Catalunya (BarcelonaTech).

5 Kolantla, D., Mikkili, S., Pendem, S.R., and Desai, A. (2020). Critical review on various inverter topologies for PV system architectures. *IET Renewable Power Generation*: 1–5.
6 Desimpelaere, K. and Schimpf, C. (2021). *Virtual Central Approach of PV String Inverters – A Cost Benefit*. KACO New Energy GmbH – A Siemens Company.
7 Spertino, F., Ahmad, J., Di Leo, P., and Ciocia, A. (2016). A method for obtaining the I–V curve of photovoltaic arrays from module voltages and its applications for MPP tracking. *Solar Energy* 139: 489–505.
8 Amiry, H., Benhmida, M., Bendaoud, R. et al. (2018). Design and implementation of a photovoltaic IV curve tracer: solar modules characterization under real operating conditions. *Energy Conversion and Management* 169: 206–216.
9 Rakesh, N., Banerjee, S., Subramaniam, S., and Babu, N. (2020). A simplified method for fault detection and identification of mismatch modules and strings in a grid-tied solar photovoltaic system. *International Journal of Emerging Electric Power Systems* 21 (4): 80–163.
10 Deutsche Gesellschaft für Sonnenenergie (DGS) (2013). *Planning and Installing Photovoltaic Systems: A Guide for Installers, Architects and Engineers*. Routledge.
11 Dupré, O. (2015). Physics of the thermal behavior of photovoltaic devices. Doctoral dissertation, Lyon, INSA, p. 159.
12 Mosalam, H.A. (2018). Experimental investigation of temperature effect on PV monocrystalline module. *International Journal of Renewable Energy Research (IJRER)* 8 (1): 365–373.
13 SMA Solar Technology AG (2013). *Central Inverter Planning of a PV Generator Planning Guidelines. SMA Company.*
14 Chen, S., Li, P., Brady, D., and Lehman, B. (2013). Determining the optimum grid-connected photovoltaic inverter size. *Solar Energy* 87: 96–116.
15 Roca Rubí, Á. (2018). Design and modelling of a large-scale PV plant. Master's thesis, Universitat Politècnica de Catalunya.
16 Berwal, A.K., Kumar, S., Kumari, N. et al. (2017). Design and analysis of rooftop grid tied 50 kW capacity solar photovoltaic (SPV) power plant. *Renewable and Sustainable Energy Reviews* 77: 1288–1299.
17 Garcia, E.E., Kimura, C., Martins, A.C. et al. (1999). TUV certificate photovoltaic solar cable resistant-40 degree 4mm 6mm PV wire PV/solar cable. *Brazilian Archives of Biology and Technology* 42 (3): 281–290.
18 Saldaña Pizarraya, G. (2020). Estudio y desarrollo de una planta fotovoltaica con paneles bifaciales. (Trabajo Fin de Grado Inédito). Universidad de Sevilla, Sevilla.
19 ABB SACE (2014). *Technical Application Papers No. 10 Photovoltaic Plants*. ABB Group.
20 Mosheer, A.D. and Gan, C.K. (2015). Optimal solar cable selection for photovoltaic systems. *International Journal of Renewable Energy Resources* 5 (2): 28–37.

21 Gremmel, H. and Kopatsch, G. (2006). *ABB Switchgear Manual*. ABB Calor Emag Schaltanlagen AG Mannheim and ABB Calor Emag Mittelspannung GmbH Ratingen.

22 Santos, R.A. and Cerqueira, S. Jr. (2017). 10 in to 2 output 1000V DC combiner box-PV strings boxes. *Journal of Microwaves, Optoelectronics and Electromagnetic Applications* 16 (1): 59–69.

23 Häberlin, H. (2012). *Photovoltaics: System Design and Practice*. Wiley.

24 Haas, H.U. (2015). fuse.on SIBA technical background information: Know-how on electrical fuses. Short circuit protection in PV systems Requirements for photovoltaic fuses. R & D SIBA GmbH & Co. KG.

25 ABB Oy low voltage products (2014). Technical Application Papers No. 10 Photovoltaic plants.

8

PV System Losses and Energy Yield

8.1 Introduction

Achieving maximum electrical energy from a PV plant is hard since the plant output energy is affected by the PV module losses. The losses are due to equipment failure and performance inefficiency. Most of the energy losses in a PV power plant are due to failures, inefficiencies, and improper design.

Successful loss reduction requires information about the sources of the loss and their relative impact on the plant performance. The task of designer is to identify the causes of losses and to provide solutions to reduce or eliminate them as much as possible. In this chapter, in reference to the technical reports and studies, various types of losses of PV power plants are discussed and the formulations to calculate the losses are presented. The performance ratio, and monthly and annual output energy productions are also discussed.

8.2 PV System Losses

PV system losses have a significant impact on the overall efficiency and output power of a PV power plant. Figure 8.1 provides an overview of various sources of energy losses in a PV plant. Figure 8.1 also shows that the losses can occur in different parts of a PV plant including the PV modules, DC to AC power conversion equipment, and the AC power transmission system. In the following sections, various types of losses in a PV plant are discussed.

8.2.1 Sunlight Losses

In this section, the types of losses related to sunlight radiation to PV modules are discussed.

Step-by-Step Design of Large-Scale Photovoltaic Power Plants, First Edition. Davood Naghaviha, Hassan Nikkhajoei, and Houshang Karimi.

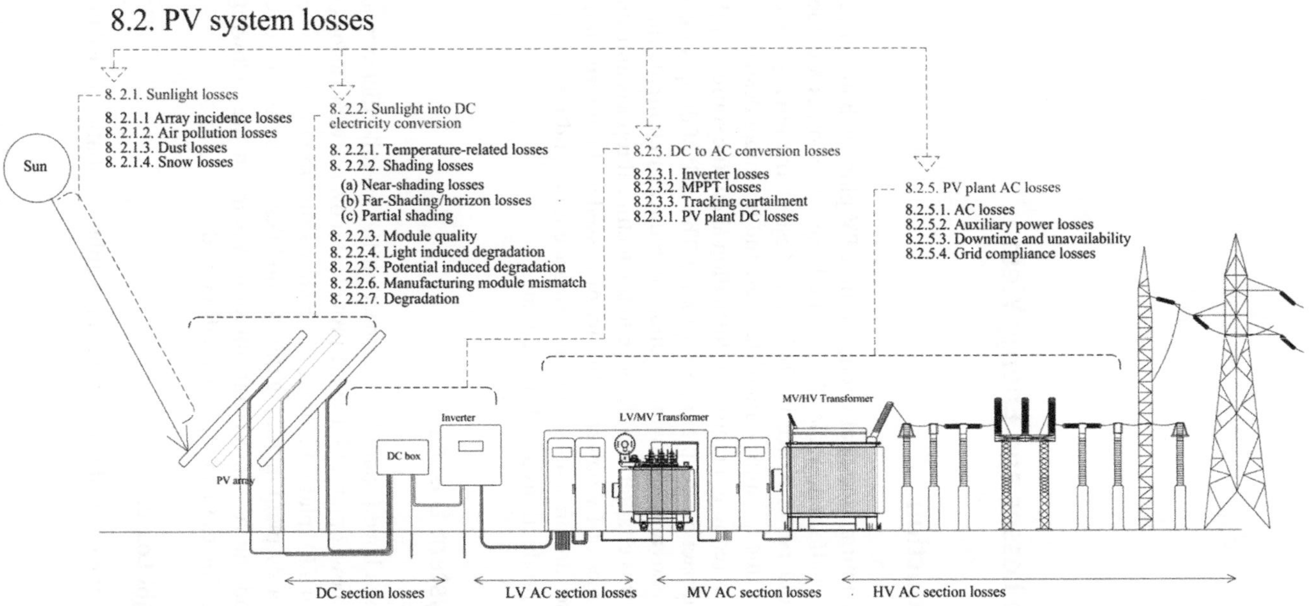
8.2. PV system losses
8. 2.1. Sunlight losses
8. 2.1.1 Array incidence losses
8. 2.1.2. Air pollution losses
8. 2.1.3. Dust losses
8. 2.1.4. Snow losses
Sun
8. 2.2. Sunlight into DC electricity conversion
8. 2.2.1. Temperature-related losses
8. 2.2.2. Shading losses
(a) Near-shading losses
(b) Far-Shading/horizon losses
(c) Partial shading
8. 2.2.3. Module quality
8. 2.2.4. Light induced degradation
8. 2.2.5. Potential induced degradation
8. 2.2.6. Manufacturing module mismatch
8. 2.2.7. Degradation
8.2.3. DC to AC conversion losses
8.2.3.1. Inverter losses
8.2.3.2. MPPT losses
8.2.3.3. Tracking curtailment
8.2.3.1. PV plant DC losses
8.2.5. PV plant AC losses
8.2.5.1. AC losses
8.2.5.2. Auxiliary power losses
8.2.5.3. Downtime and unavailability
8.2.5.4. Grid compliance losses
PV array
DC box
Inverter
LV/MV Transformer
MV/HV Transformer
DC section losses
LV AC section losses
MV AC section losses
HV AC section losses

Figure 8.1 An overview of various sources of energy losses in a PV plant.

8.2.1.1 Array Incidence Losses

One of the important parameters for estimating the DC output power of the modules is to consider the optical losses. Direct rays of sunlight are not always perpendicular to the module plane, which is one of the causes of light losses. By changing the angle of direct sunlight radiation to the module plane, which is called the angle of incidence (AOI), as shown in Figure 8.2, the optical losses can be minimized. For the flat panel modules, light losses are related to the light reflection loss from the glass surface in front of the module. The light reflection for an AOI greater than 60° is significantly increased as shown in Figure 8.3. The result of this type of loss is less radiation reaching the cells inside the module, which in turn reduces the module energy production [3].

The angular losses (AL) of a PV module is calculated from the amount of normal radiation and cleanliness of the module surface from (8.1) [4].

$$\mathrm{AL}(x)=1-\frac{T(x)}{(T0)}=1-\frac{1-R(x)-A(x)}{1-R(0)-A(0)}\cong 1-\frac{1-R(x)}{1-R(0)} \tag{8.1}$$

In (8.1), $R(x)$, $A(x)$, and $T(x)$ are, respectively, the reflectance, transmittance, and absorptance at the angle x.

In [5], the optical losses of a 1.2 MW PV power plant located in Spain due to dust and AOI are evaluated. The results of this study show that the optical losses due to AOI reach 1% when the PV plant is equipped with a solar tracking system. The optical losses are between 2 and 3% for the fixed angle system in summer and 8% in certain winter months.

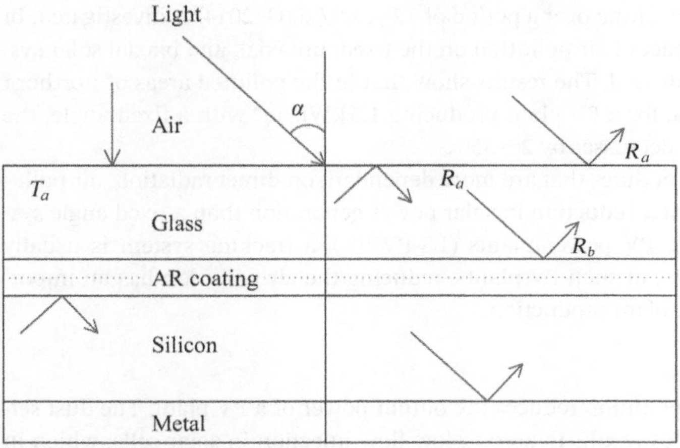
Light
α
Air
R_a
T_a
R_a
Glass
R_b
AR coating
Silicon
Metal

Figure 8.2 Schematic diagram of reflections and transmissions that contribute to the final reflectance of a crystalline silicon PV module. *Source:* Modified from Martin and Ruiz [1].

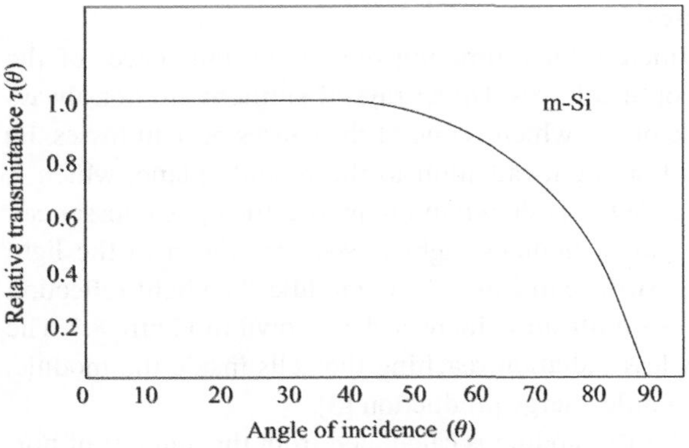
Relative transmittance τ(θ)
1.0
0.8
0.6
0.4
0.2
m-Si
0
10
20
30
40
50
60
70
80
90
Angle of incidence (θ)

Figure 8.3 Angular transmittance for a particular example of a c-Si module. *Source:* Modified from Shah [2].

8.2.1.2 Soiling Losses

Many environmental factors affect the production of PV power plants, and air pollution is one of the main reasons for the decline in energy production. Sunlight can be transmitted, absorbed, or diffused as it passes through the atmosphere. Air pollution causes the scattering and absorption of sunlight. The air pollution is created by fossil fuel sources and industrial pollution, and reduces the net radiation to PV modules. The study of a PV plant with 119 modules in China shows that solar power generation capacity has decreased by an average of 11–15% between 1960 and 2015 due to air pollution and the resulting reduced radiation on the surface of PV modules [6].

In [7], the impact of air pollution on the production of PV power plants in several locations of China over a period of 12 years (2003–2014) is investigated. In this study, the impact of air pollution on the fixed, uniaxial, and biaxial solar systems has been evaluated. The results show that in the polluted areas of northern and eastern China, for a PV plant producing 1.5 kWh/m^2 with a fixed angle, the power production decreases by 25–35%.

For tracked PV modules that are more dependent on direct radiation, air pollution leads to a greater reduction in solar power generation than a fixed angle system. In large-scale PV power plants (LS-PVPPs), a tracking system is usually installed. Therefore, in such PV plants, reducing the air pollution has an important impact on the plant production.

8.2.1.3 Dust Losses

Dust as a natural pollution reduces the output power of a PV plant. The dust settling on PV modules results in zero or low flow injection in solar cells, which in turn leads to reduced module power generation as shown in Figure 8.4. Furthermore, the dust produces an electric charge and provides a path for the

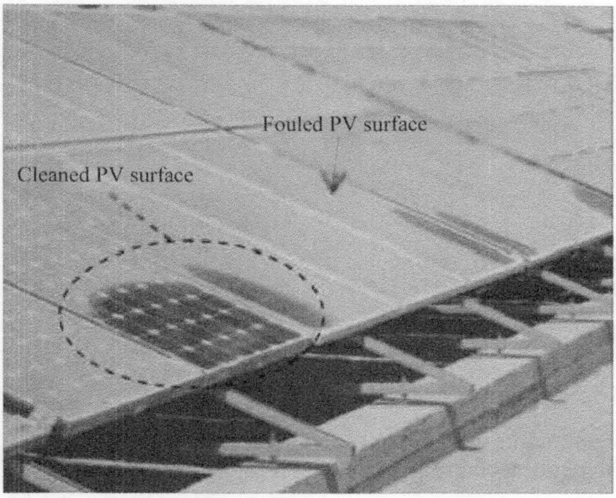
Fouled PV surface
Cleaned PV surface

Figure 8.4 Accumulation of sand particle on PV panels. *Source:* Abd-Elhady et al. [8], figure 01/Heat Transfer Research, Inc.

electric current produced by the PV modules. The current circulation is converted into heat, which, if significant, can lead to hot spots in the long run [9].

The results of an investigation on a large-scale PV plant located in the United States show that dust rates range from 0 to 0.3% per day [10]. The dust losses are reported at an average annual rate of 4% for the PV plants located in the southwestern of the United States. The results also show that without regular PV modules cleaning, the output plant energy is reduced by 20–40%. The rate of natural pollution varies from region to region. In some PV plant sites, the panels need to be washed twice a week, while in some other sites, daily washing may be required. Considering a 4% loss for the total solar energy produced in the world, which is more than 270 GW, the total losses reach 10 GW [10].

The effect of dust on the efficiency of a PV power plant depends on the following parameters:

- Particle thickness
- Particle density
- Particle type

There are 17 types of dusts including fine dust, sand, limestone, carbon, and clay [10]. In [11], the effect of three types of dust including limestone, ash, and red soil on the efficiency of PV plants are compared, Figure 8.5. Figure 8.6 shows the changes in the efficiency of PV plants in terms of the dust particle density. Based on the results of the study presented in [9], the impact of red soil on reducing the plant energy production is much greater than those of the limestone and ash.

Limestone
Ash
Red soil

Figure 8.5 The three types of pollutants used in the experiment. *Source:* Kaldellis and Kapsali [9], figure 02 (p.02)/with permission from Elsevier.

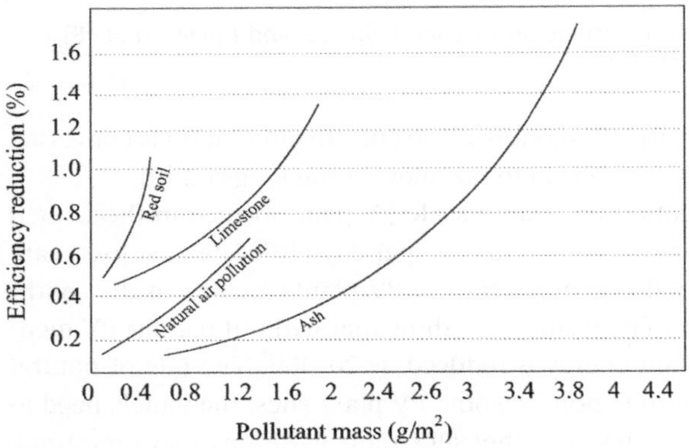
Efficiency reduction (%)
1.6
1.4
1.2
1.0
0.8
0.6
0.4
0.2
Red soil
Limestone
Natural air pollution
Ash
0
0.4
0.8
1.2
1.6
2
2.4
2.6
3
3.4
3.8
4
4.4
Pollutant mass (g/m²)

Figure 8.6 Efficiency differences between the clean and the polluted pair panels for various mass depositions in cases of naturally and artificially polluted PV panels. *Source:* Modified from Kaldellis and Kapsali [9].

Another factor influencing the impact of dust on PV plant generation is the type of module used in the strings. The results of a study of three types of modules used in Thailand show that the rate of reduction of electrical energy in one month for the a-Si, m-Si, and p-Si modules is equal to 3.5, 2.96, and 2.83%, respectively [9].

8.2.1.4 Snow Losses

To predict the performance of a PV plant in cold regions experiencing snowfall, it is important to consider the effect of snow on the plant performance. It should be noted that in the areas with heavy annual snowfall, a larger tilt angle should be selected so that the snow is directed downwards faster.

The study carried out in [12] provides the results on the amount of snowfall in various US cities. The average percentage of annual snowfall for slope angles of 20° and different latitudes has been given.

8.2.2 Sunlight into DC Electricity Conversion

8.2.2.1 Temperature-Related Losses

One of the parameters affecting the efficiency of a PV module is the temperature of solar cells. As the temperature of the cells increases, the open circuit voltage of the PV module decreases. This reduces the maximum power that can be received from the module. The relationship between the open-circuit voltage of the PV module and the temperature is expressed by (8.2) and (8.3),

$$V_{oc} = \frac{kT}{q}\ln\left(\frac{I_{sc}}{I_0}+1\right) \tag{8.2}$$

$$I_0 = CT^3\exp\left(-\frac{E_{g0}}{kT}\right) \tag{8.3}$$

where C is a temperature-independent constant, k is the Boltzmann constant, E_{g0} is the band gap of the material extrapolated to absolute zero temperature, T is the temperature, q is the elementary charge, I_{sc} is the short-circuit current, and I_0 is the diode saturation current. As irradiance decreases, I_{sc} decreases linearly, and consequently, the efficiency of the PV modules is reduced due to the decrease in V_{oc} with respect to I_{sc}. Equations (8.2) and (8.3) indicate that the exponential dependence of I_0 on the temperature has a greater effect on the open-circuit voltage than the linear dependence of V_{oc} on the T parameter. The open-circuit voltage decreases when the temperature increases [13].

To measure the temperature losses of a PV power plant, the operating temperature of the cells of the PV modules must be specified. This temperature cannot be measured directly because the cells are not accessible from the outside. The outside surface temperature of the PV modules can be measured directly. However, the cell temperature depending on its structure is usually 1–3° higher than its outside surface temperature.

Methods such as open-circuit voltage measurement of PV modules have been proposed to determine the cell temperature. However, due to its difficulty, the models that express the relationship between cell temperature and PV module installation configuration, module type, and environmental parameters are used. There are 20 models for estimating the temperature of a PV module as presented in [14].

In [15], the effect of temperature on annual energy production for various technologies of PV modules installed in Cyprus is analyzed. The results of this study show that the average annual loss of DC energy production is equal to 9% for multi-crystalline silicon technology, 8% for monocrystalline silicon technology, and 5% and for thin-film technology.

8.2.2.2 Shading Losses

Shading losses are divided into three main categories as shown in Figure 8.7: near shading, far shading, and electrical shading losses.

a) **Near-Shading Losses**

One of the most important causes of losses for large-scale solar power plants is the occurrence of near shading. Near shading can happen due to the adjacent buildings, chimneys, air vents, trees, spaces adjacent to the roof, overhead lines, power pole base, front strings, and other shading objects. In Chapter 5, the details of string shading calculations for the fixed, single-axis, and two-axis PV power plant were presented.

For accurate modeling of near-shadow losses, it is recommended that a three-dimensional view of the tall objects and near-shadow obstacles be created in the modeling software. The near-shading losses can be significantly large. Therefore, accurate modeling is greatly important. The use of filament inverters instead of central inverters is one way to reduce the impact of shadow reduction on the PV modules performance.

b) **Far-Shading/Horizon Losses**

Mountains and hills that are far from a PV plant can block the sunlight and shade the PV plant. To calculate the shading losses caused by distant mountains on a PV plant, a horizon line, which includes the azimuth points and altitude, should be drawn as shown in Figure 8.8. In the example shown in Figure 8.8, in a distance there is a horizon line of a mountain or a high hill to the east of the PV plant.

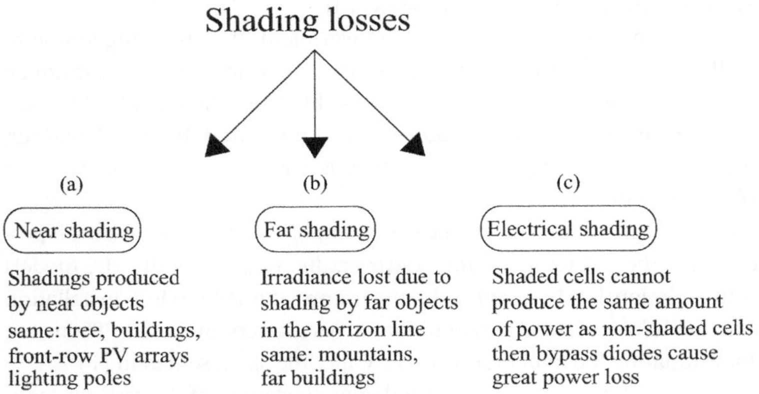
Shading losses
(a)
(b)
(c)
Near shading
Far shading
Electrical shading
Shadings produced by near objects same: tree, buildings, front-row PV arrays lighting poles
Irradiance lost due to shading by far objects in the horizon line same: mountains, far buildings
Shaded cells cannot produce the same amount of power as non-shaded cells then bypass diodes cause great power loss

Figure 8.7 Different categories of shading losses.

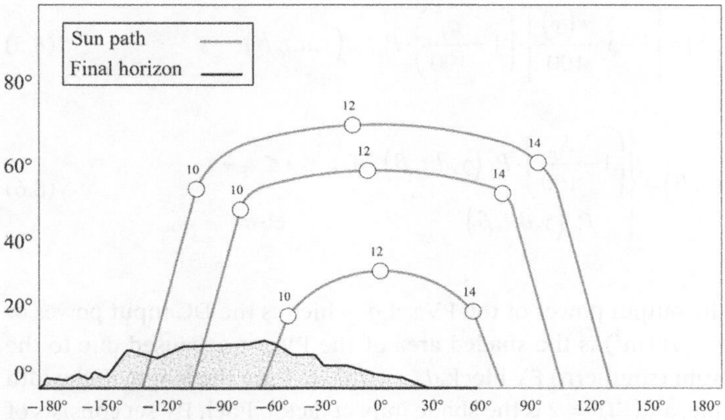
Sun path
Final horizon
80°
60°
40°
20°
0°
12
14
10
12
14
10
12
10
14
−180°
−150°
−120°
−90°
−60°
−30°
0°
30°
60°
90°
120°
150°
180°

Figure 8.8 Sample of a horizon line.

This can be seen in the graph of the path of the sun throughout the year, which is behind the horizon in some months. A part of the sunlight does not reach PV modules and causes shading losses.

To obtain the horizon line, software such as SAM, CARNAVAL, SOLMETRIC, HORIZON, METEONORM are used. In some cases, the information is manually inputted into the software such as PVSYST to accurately measure the distant shading losses. The formulation to calculate the shading losses are presented in [16].

c) **Partial Shading**

The effect of partial shadows on the electrical production of PV power plants is nonlinear. This effect is modeled by partitioning the module strands. Partial shading causes two types of losses in a PV module [17]:

- Losses due to reduced radiation on the module surface
- Losses due to connection of the shaded and shadowless modules with different I–V characteristics

A module installed in the horizontal configuration with respect to the equator, usually experiences less electrical shadow losses than the modules installed in the vertical configuration due to the connection of diodes. On average, the partial shading losses are between 20–25% [18].

Considering the shadow effect of the front strings, the DC input power of the inverter for a large-scale PV plant is given as [19]

$$P_{\text{in}}(q) = \left[1 - \frac{A_{s,l}(q)}{N_s \cdot N_p \cdot L_{\text{pv},1} \cdot L_{\text{pv},2}} \cdot \text{SIF}\right] \cdot \left(1 - \frac{\eta_{l,\text{dc}}}{100} \cdot \text{PL}_{\text{DC}}\right) \cdot N_s \cdot N_p \cdot \eta_{\text{mppt}}\left(N_s \cdot N_p \cdot P_{\text{pv}}(y,d,t,\beta)\right) \cdot P_{\text{pv}}(y,d,t,\beta) \quad (8.4)$$

$$P_{pv}(y,d,t,\beta) = \left[1 - y \cdot \frac{r(y)}{100}\right] \cdot \left(1 - \frac{d_f}{100}\right) \cdot P_{m-sh}(y,d,t,\beta) \tag{8.5}$$

$$P_{m-sh}(y,d,t,\beta) = \begin{cases} \left(1 - \frac{S_p}{100}\right) \cdot P_m(y,d,t,\beta) & \text{if } t_1 \le t \le t_1 + t_2 \\ P_m(y,d,t,\beta) & \text{else} \end{cases} \tag{8.6}$$

where $P_{in}(q)$ is the output power of the PV set q, which is the DC input power of each inverter. $A_{s,f}(q)$ (m^2) is the shaded area of the PV set q, caused due to the shading by the front (southern) PV block, $L_{pv,1}$ and $L_{pv,2}$ are the length and width of each PV module, and SIF = 2 is the shade impact factor. Each PV set consists of N_p PV strings ($N_p \geq 1$), while each string comprises N_s PV modules that are connected in series ($N_s \geq 1$). $\eta_{l,DC}$ is the DC cables power loss coefficient. PL_{DC} is the power-length product in kWp·m of the DC cables from PV modules to DC/AC inverters. η_{mppt} is the MPPT efficiency of the inverter. $P_{pv}(y, d, t, \beta)$ is the actual output power of each PV module on year "y" ($1 \leq y \leq n$), day d ($1 \leq d \leq 365$), and at time t ($1 \leq t \leq 24$). $r(y)$ (%/year) denotes the annual reduction coefficient of the PV module output power. If $y = 1$, then $r(y) = 0$. For $1 < y \leq n$, the value of r is specified by the PV module manufacturer. d_f (%) is the PV module output power derating factor due to the dirt deposited on the module surface. The derating can reach 6.9% for large-scale PV plants. $P_{m-sh}(y, d, t, \beta)$ (kW) is the output power of each PV module at the maximum power points (MPP), which is calculated by considering the shading condition.

A PV plant designer should specify the percentage of the PV module total area that is shadowed by the surrounding obstacles, i.e., S_p (%), the time of the day, i.e., t_1 (h), that these obstacles start to shadow the PV modules, and the corresponding time duration of the shadow t_2 (h). $P_m(y, d, t, \beta)$ (kW) is the power that is produced by each PV module at the MPP. The parameter β is the PV modules tilt angle ($0 \leq \beta \leq 90$).

In [18], the losses of three types of solar modules included in a single string, short parallel strings, and multi-strings are investigated for both static and dynamic shading scenarios, see also Figure 8.9. In the static shading scenario, two levels of radiation, and in the dynamic shading scenario, multiple levels of radiation are considered to simulate the shadow impacts (Figure 8.10). The P–V characteristics of three types of PV modules are presented for the static and dynamic shading scenarios in Figures 8.11 and 8.12. The results show that the characteristic of PV modules with the occurrence of shadows becomes more complex than that of a normal state. This complexity is due to the formation of several MPPs in the characteristic curve of the PV power plant. The results of study presented in [18] show that with 25% partial shading, all connection

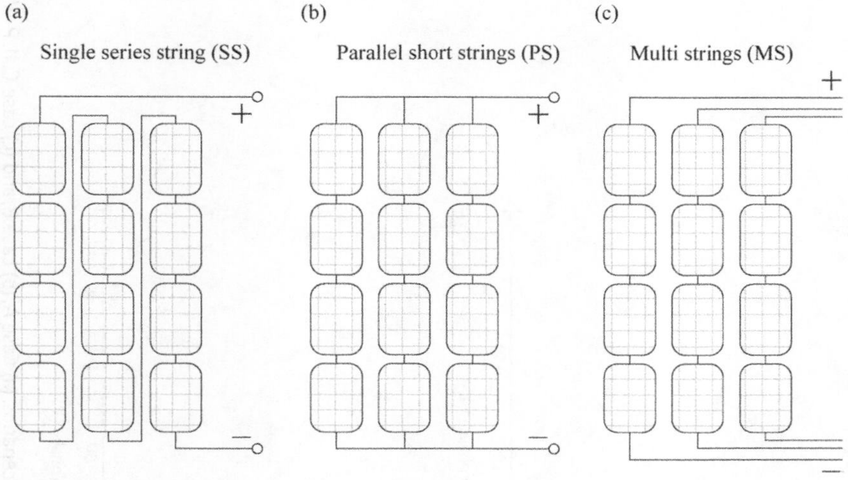
(a)
Single series string (SS)
+
−
(b)
Parallel short strings (PS)
+
−
(c)
Multi strings (MS)
+
−

Figure 8.9 Different PV power generators considered in the study. (a) Single-string generator, (b) parallel short-string generator, and (c) multi-strings generator. *Source:* Modified from Satpathy and Sharma [18].

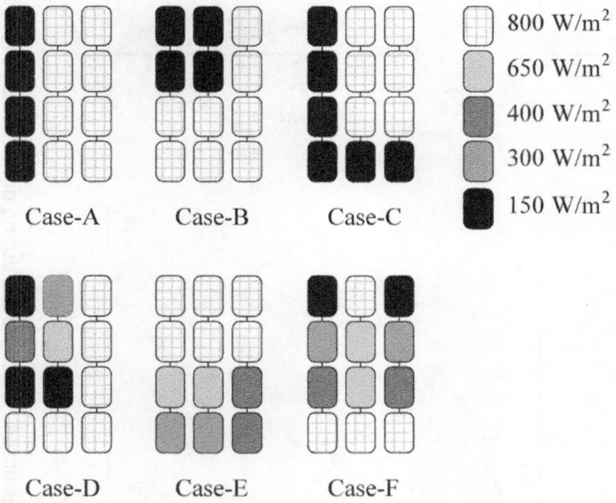
800 W/m2
650 W/m2
400 W/m2
300 W/m2
150 W/m2
Case-A
Case-B
Case-C
Case-D
Case-E
Case-F

Figure 8.10 Static random and dynamic shading scenarios. *Source:* Modified from Satpathy and Sharma [18].

arrangements have almost the same losses, i.e. 50%. However, as the shading increases to 50%, the losses of multi-string connection arrangements show less losses (around 31.63%). Therefore, to increase the annual energy production of a PV plant exposed to partial shading, the use of series and parallel arrangements should be minimized.

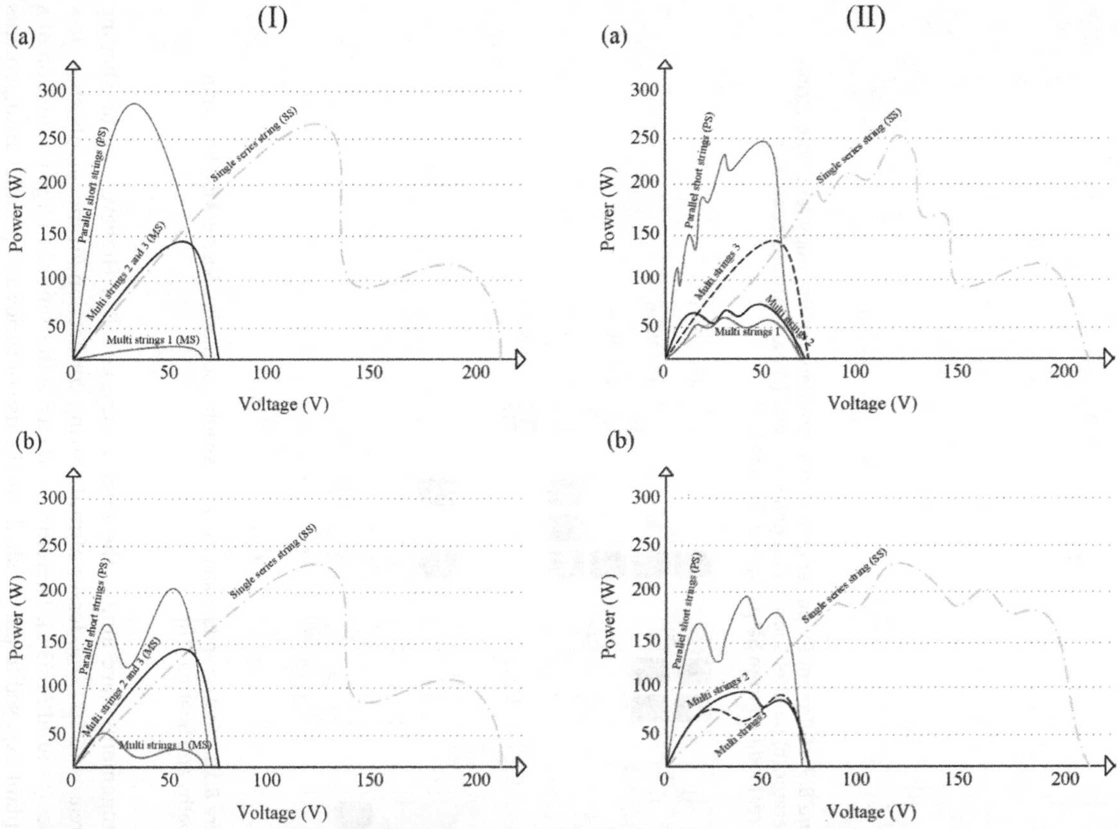
(I)
(a)
Power (W)
Voltage (V)
Parallel short strings (PS)
Multi strings 2 and 3 (MS)
Single series string (SS)
Multi strings 1 (MS)
(b)
Power (W)
Voltage (V)
Parallel short strings (PS)
Multi strings 2 and 3 (MS)
Single series string (SS)
Multi strings 1 (MS)
(II)
(a)
Power (W)
Voltage (V)
Parallel short strings (PS)
Multi strings 3
Single series string (SS)
Multi strings 1
Multi strings 2
(b)
Power (W)
Voltage (V)
Parallel short strings (PS)
Multi strings 2
Multi strings 3
Single series string (SS)

Figure 8.11 I. P–V curves of the PV generators during different static random shading scenarios. (a) Case A, (b) case B, and (c) case C. II. P–V curves of the PV generators during different dynamic shading scenarios. (a) Case D, (b) case E, and (c) case F. *Source:* Modified from Satpathy and Sharma. [18].

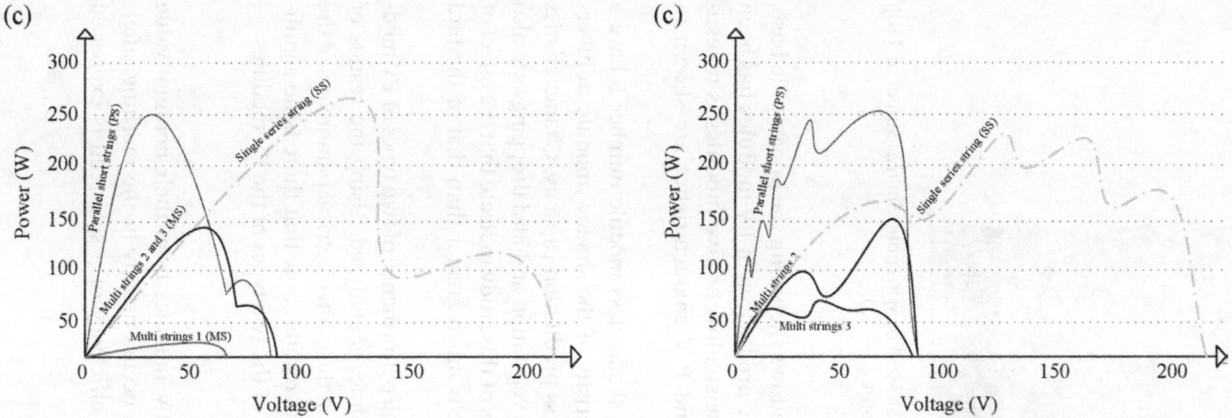
(c)
Power (W)
300
250
200
150
100
50
0
50
100
150
200
Voltage (V)
Parallel short strings (PS)
Multi strings 2 and 3 (MS)
Single series string (SS)
Multi strings 1 (MS)
(c)
Power (W)
300
250
200
150
100
50
0
50
100
150
200
Voltage (V)
Parallel short strings (PS)
Multi strings 2
Multi strings 3
Single series string (SS)

Figure 8.11 (Continued)

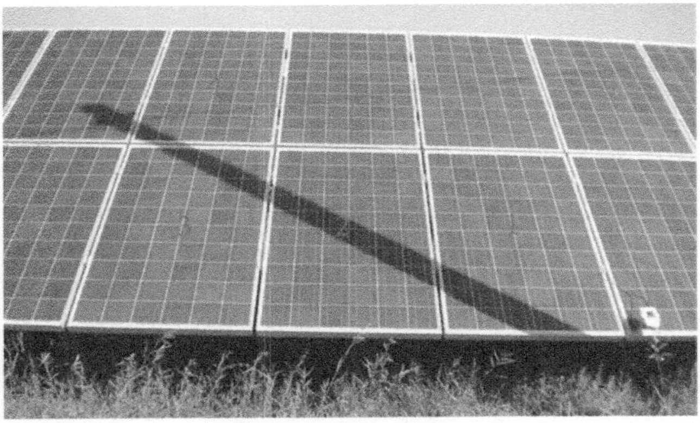

Figure 8.12 The numbers of the PV modules shaded by wire pole. *Source:* Sun et al. [20], figure 04 (p. 05)/Hindawi Publishing Corporation/CC BY 3.0.

In [20], the effect of three types of shadows including the power pole base, the vegetation, and bird droppings on the production of PV modules has been investigated. The effect of the beam base shadow on five modules is investigated based on Figure 8.12. The I–V and P–V characteristics are shown in Figure 8.13.

The P–V and I–V characteristics of the shadowless module number 5 have a maximum power point, and the characteristic of the shadow module numbers 1–4 have several maximum points with less production capacity. Characteristics of shadowless and shaded modules due to vegetation and bird droppings are also shown in Figures 8.14 and 8.15. The results of this study indicate that the effect of vegetation shade and electricity pole base is much greater than that of the bird droppings.

In [21], the effect of partial shading on the performance of two types of PV modules, i.e. polycrystalline and monocrystalline, is analyzed. Using the results of measurement tests, the impact of shadow on the main electrical parameters of the PV modules has been investigated, and the results show that there is no significant difference between the performances of the two types of the PV modules.

8.2.2.3 Low Irradiance

Low radiation reduces the efficiency of PV modules from their nominal value under the STC conditions. Low radiation occurs usually in the morning, afternoon, and cloudy days. The normalized efficiency of a solar cell is expressed as [22].

$$\eta_{\text{rel}}(G,T) = \frac{\eta}{\eta_{\text{STC}}} = \frac{P_{\max}(G,T)}{P_{\max,\text{STC}}} \frac{G_{\text{STC}}}{G} \tag{8.7}$$

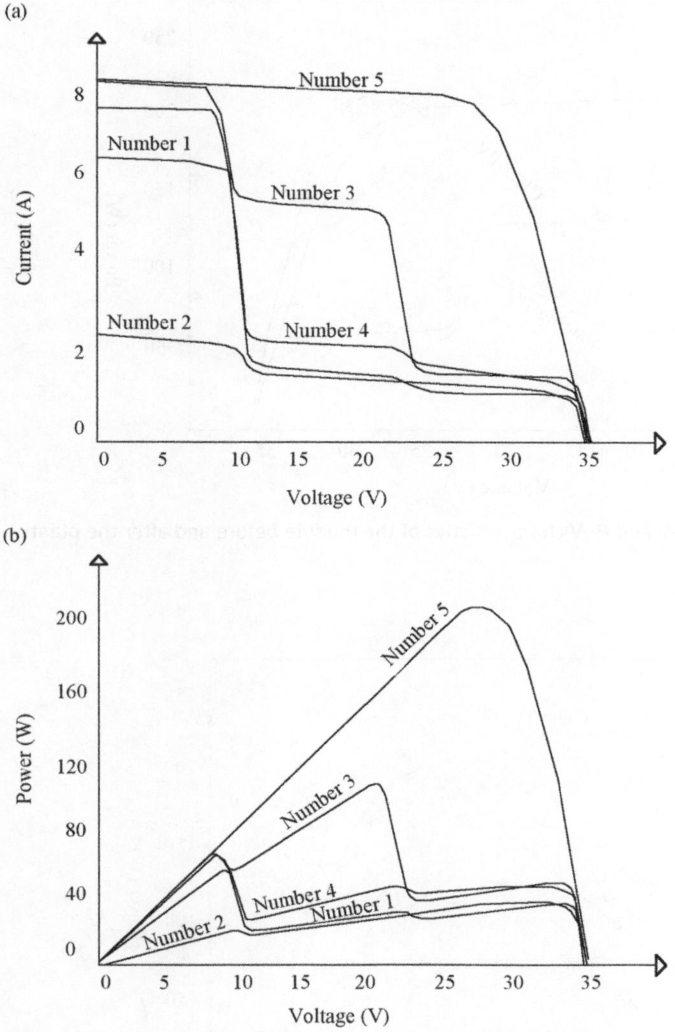
(a)
Current (A)
Voltage (V)
Number 5
Number 1
Number 3
Number 2
Number 4
(b)
Power (W)
Voltage (V)
Number 5
Number 3
Number 4
Number 1
Number 2

Figure 8.13 I–V curves and P–V curves of the five shaded PV modules. (a) I–V curves; (b) P–V curves [20].

where $P_{max,\, STC}$ is the maximum output power at the STC conditions, $P_{max}(G, T)$ is the maximum output power at a given irradiance G and temperature T, and

$$G_{STC} = 1000\frac{W}{m^2}.$$

In [23], the variations of efficiency with respect to the radiation for two types of PV modules, i.e. CdTe and c-Si, are studied. The study considers four days in winter

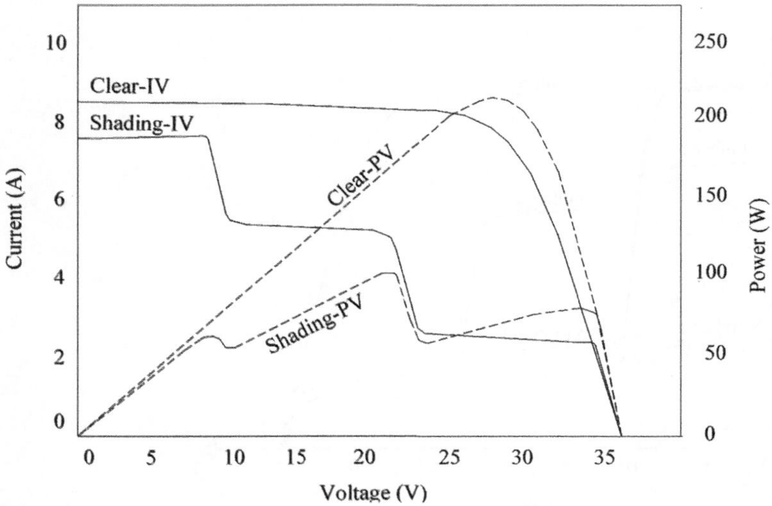

10
8
6
4
2
0
Current (A)
Clear-IV
Shading-IV
Clear-PV
Shading-PV
250
200
150
100
50
0
Power (W)
0
5
10
15
20
25
30
35
Voltage (V)

Figure 8.14 The I–V and P–V characteristics of the module before and after the plant removal [20].

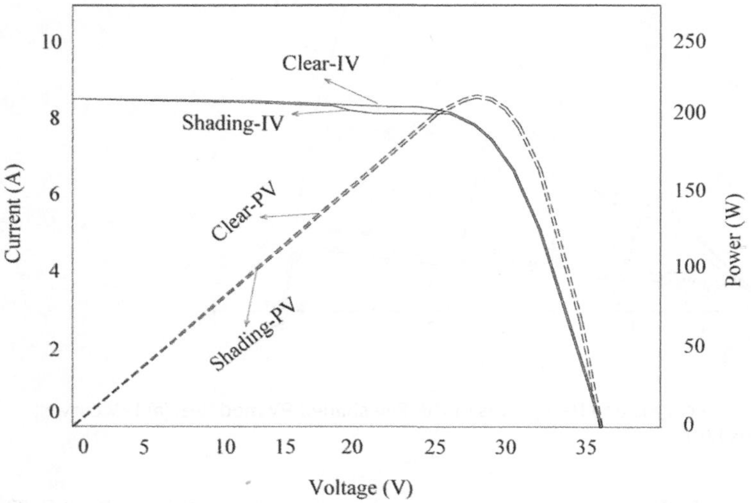

10
8
6
4
2
0
Current (A)
Clear-IV
Shading-IV
Clear-PV
Shading-PV
250
200
150
100
50
0
Power (W)
0
5
10
15
20
25
30
35
Voltage (V)

Figure 8.15 The I–V and P–V characteristics of the module before and after the bird droppings removal [20].

and four days in summer. The results show that by reducing the radiation from 1000 to 500 W/m^2 in summer, the efficiency decreases from 98 to 94.5%. Moreover, for radiations less than 500 W/m^2, the efficiency of CdTe type PV module is higher than that of the c-Si type. The reason is that the maximum spectral response of the CdTe module is in the visible region where photons are absorbed with high energy over their wavelengths. The maximum spectral response of the c-Si PV module is located in the infrared region, where photons are absorbed with low energy over their wavelengths.

In large-scale PV plants, most shadings are caused by moving clouds, which make the radiation become needle-shaped as shown in Figure 8.16. This creates losses, fluctuations in the PV plant production, and possible interruption of the solar tracking system.

In [24], the effects of the PV modules connection arrangement and their direction on the losses due to moving clouds have been investigated. In this study, three connection arrangements of PV modules are considered as shown in Figure 8.17. The results show that as the number of PV modules per string increases, the losses due to moving clouds increase too, Figure 8.18. This is due to the fact that the radiation difference between the PV modules is higher for longer modules.

The study presented in [24] shows that the number of parallel strings has a small effect on the losses. However, the array orientation angle has a significant effect on the losses caused by moving clouds. This is due to the heterogeneous distribution of the clouds shadow edges. The study also provides the losses of six types of PV modules in terms of changes in the array orientation angle. It is shown that when the direction angle is 90°, the losses caused by moving clouds are minimum.

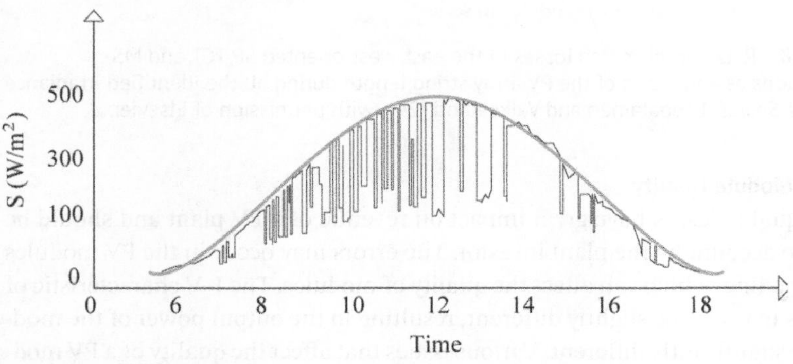
S (W/m²)
500
300
100
0
0
6
8
10
12
14
16
18
Time

Figure 8.16 High temporal variability of solar irradiance caused by moving cloud shadows [23].

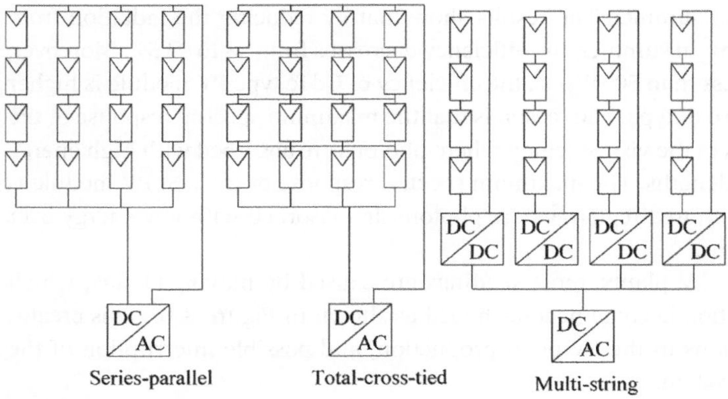
DC
AC
Series-parallel
Total-cross-tied
Multi-string
DC
DC

Figure 8.17 The electrical connections for the studied PV array configurations (series parallel [SP], total cross tied [TCT], and multi-string [MS]). *Source:* Lappalainen and Valkealahti [24] /with permission of Elsevier.

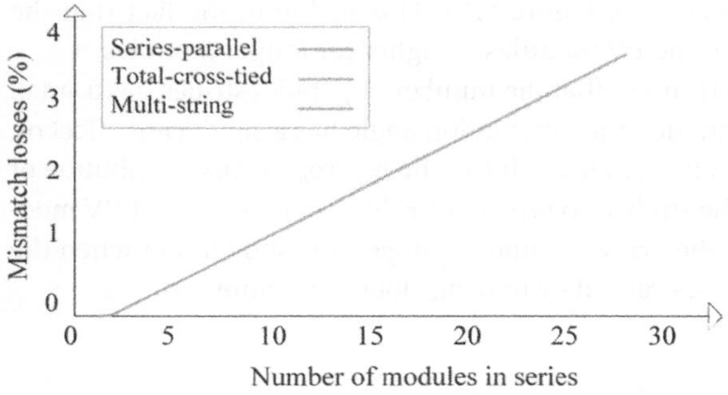
Mismatch losses (%)
Number of modules in series
Series-parallel
Total-cross-tied
Multi-string
0
1
2
3
4
0
5
10
15
20
25
30

Figure 8.18 Relative mismatch losses of the east-west-oriented SP, TCT, and MS configurations as a function of the PV array string length during all the identified irradiance transitions. *Source:* Lappalainen and Valkealahti [24] /with permission of Elsevier.

8.2.2.4 Module Quality

Module quality losses have great impact on revenue of a PV plant and should be taken into account by the plant investor. The errors may occur in the PV modules production line, which can affect the quality of modules. The I–V characteristic of solar cells may also be slightly different, resulting in the output power of the modules to be significantly different. Various issues that affect the quality of a PV module are discussed in the sequel.

8.2.2.5 Light-Induced Degradation

Light-induced degradation (LID) loss is a defect that occurs in crystal modules in the first hours of sunlight and affects the actual performance of the module. These losses can be caused by the presence of boron, oxygen, or other chemicals left over from the etching process of cell production. LID losses affect only common boron P-type wafers and are related to the quality of wafer production. These losses can range from 1 to more than 2% of the rated module power [25].

8.2.2.6 Potential-Induced Degradation

Potential-induced degradation (PID) is a performance degradation mechanism, which occurs in a PV module due to stray currents. The PID causes gradual losses of power and can go up to more than 30%. The PID generally occurs in PV modules with ungrounded inverters. Two issues are resulted from the PID: (i) loss of useful module power; and (ii) degradation of the module front surface passivation, which results in increased recombination and cell damages. The PID usually begins a few years after the installation of a PV module.

In a PV string with 15–20 modules connected in series in order to raise the DC voltage, some cells in the end-string modules have a large potential difference (ΔV) with respect to the module frame, which is grounded. This voltage difference ΔV can cause electrons from the PV cells to go free and discharge through the grounded frame. This phenomenon leads to a leakage current that flows through the module insulation and glass.

Various paths for the leakage current from cells to the grounded frame are shown in Figure 8.19a. The PID causes a drop in the shunt resistance R_{sh} of a PV module. This reduces the maximum generated power and open-circuit voltage, leading to a reduction in the fill-factor and cell efficiency, Figure 8.19b. The impact of PID is accelerated at higher temperatures and when the top glass becomes wet and conductive during high humidity conditions [26].

8.2.2.7 Manufacturing Module Mismatch

One of the reasons for power losses in PV power plants is the mismatch that happens during the construction of single-string modules. This mismatch occurs when modules with different current–voltage characteristics are connected to

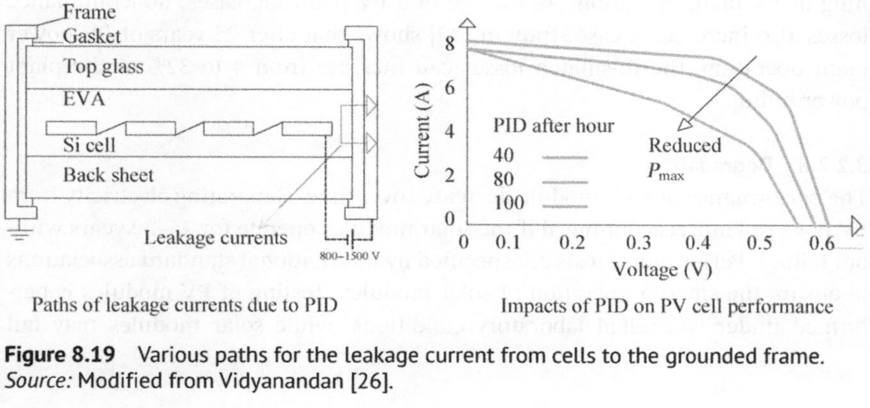
Frame
Gasket
Top glass
EVA
Si cell
Back sheet
Leakage currents
800–1500 V
Paths of leakage currents due to PID
Current (A)
8
6
4
2
0
PID after hour
40
80
100
Reduced
P_{max}
0
0.1
0.2
0.3
0.4
0.5
0.6
Voltage (V)
Impacts of PID on PV cell performance

Figure 8.19 Various paths for the leakage current from cells to the grounded frame. *Source:* Modified from Vidyanandan [26].

each other. In this case, the output power of the string is less than the total power of the string modules.

Despite the efforts of manufacturers, there are always mismatches in the electrical parameters of PV modules, even with the same rated specifications. Normally, the maximum output power of crystalline silicon (C-Si) modules with an accuracy of 5% is given in the datasheet. It indicates the potential difference in the I–V characteristic of the modules that occurs during the manufacturing process. For an accuracy of 5%, the mismatch losses of a module can reach 0.5% in the STC conditions [27].

The I–V and P–V characteristics of a solar cell provide five main parameters V_{oc}, V_{mpp}, I_{sc}, I_{mpp}, and P_{mpp}. In [28], the probability density function of these parameters is provided for 105 modules with the same nominal characteristics. The parameters are obtained using measurement data. A mismatch in the voltage parameter of a module is not important for PV plant designers. The asymmetric distribution of the module flow parameter should be considered in the analysis of module construction losses [28]. If a string needs to provide MPP current, the modules with the same MPP current are stacked together in the string. With this arrangement, the module mismatch losses can be reduced to below 0.1% [27].

There are construction mismatch losses, which depend on the environmental conditions such as radiation, temperature, and so on. For example, increasing the radiation will result in an increase in the mismatch losses. Moreover, if the radiation falls below a threshold, the mismatch losses can even be negative. This corresponds to an increased productivity [29].

In [29], the mismatch losses for a 1 MW PV power plant consisting of 4128 modules operating under various environmental conditions are calculated. The results are shown in Table 8.1 and Figure 8.20. In this study, the mismatch losses under STC conditions are 0.35%. It is shown that increasing radiation will increase the losses. When the radiation is less than 200 W/m^2, the percentage of radiation losses is negative. Although the percentage of noncompliance losses of modules construction can be reduced to below 0.5%, this reduction is achieved at the beginning of PV plant operation. As the age of a PV plant increases, noncompliance losses also increase. A case study in [30] shows that after 25 years of PV power plant operation, the mismatch losses can increase from 4 to 32% of the plant power rating.

8.2.2.8 Degradation

The performance of a PV module decreases over time. Generating electricity from PV power plants is economical if the solar modules operate for 25–30 years without failure. Performance tests are specified by international standard associations to ensure the smooth operation of solar modules. Testing of PV modules is performed under controlled laboratory conditions, while solar modules may fail

Table 8.1 Solar irradiance and cell temperatures [29].

Working condition	Solar irradiance (p.u.-W/m^2)	Cell temperature (°C)
A	0.05–50	16.5
B	0.1–100	18
C	0.2–200	21
D	0.3–300	24
E	0.5–500	30
F	1–1000	45

Source: Based on Pavan et al. [29].

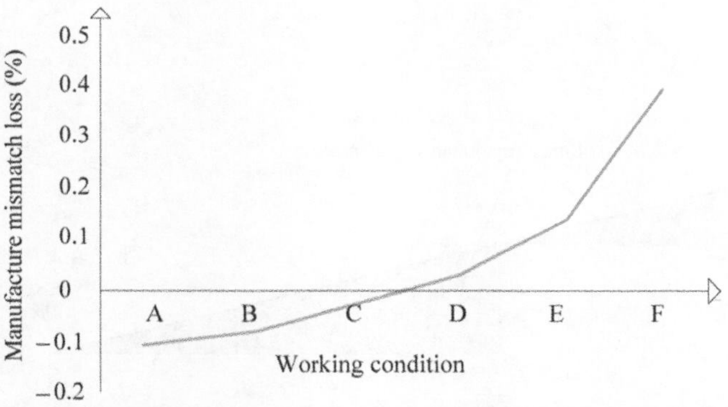
Manufacture mismatch loss (%)
0.5
0.4
0.3
0.2
0.1
0
−0.1
−0.2
A
B
C
D
E
F
Working condition

Figure 8.20 Manufacturing mismatch losses at different working conditions. *Source:* Modified from Pavan et al. [29]/with permission of Elsevier.

prematurely under certain environmental conditions such as the temperature and/or humidity increase. The module failure causes energy efficiency losses; hence, it is necessary to evaluate the performance of PV modules under realistic operating conditions.

The conditions that can cause failure of PV modules include [31]:

- degradation of the packaging materials;
- loss of adhesion;
- degradation of the cell/module interconnects;
- degradation due to moisture intrusion;
- degradation of the semiconductor device.

The results presented in the literature show that when PV modules are placed outdoors, their output power is affected by the linear failure rate of the modules over time. The extent of this failure depends on the PV module manufacturing technology.

The failure rate for crystalline silicon PV modules is much lower than that of amorphous Si and CIS PV modules. Considering the crystalline silicon as a reference, the linear rate of annual power reduction varies from 0.3 to more than 1% [32].

Manufacturers of PV modules typically guarantee a module life of 25 years. As shown in Figure 8.21, the warranty curve typically promises that the modules produce at least 80% of the rated capacity in 25 years. The module damages may be chemical, electrical, thermal, and/or mechanical.

PV modules usually degrade more rapidly in the first few years of life. In general, the output power of a PV module typically decreases by about 0.5%/year. Table 8.2 shows the average annual production decline that has been reported for PV modules with various technologies since 2000. According to the table, the thin film PV modules (a-Si, CdTe, and CIGS) degrade faster than the Si crystal modules [20].

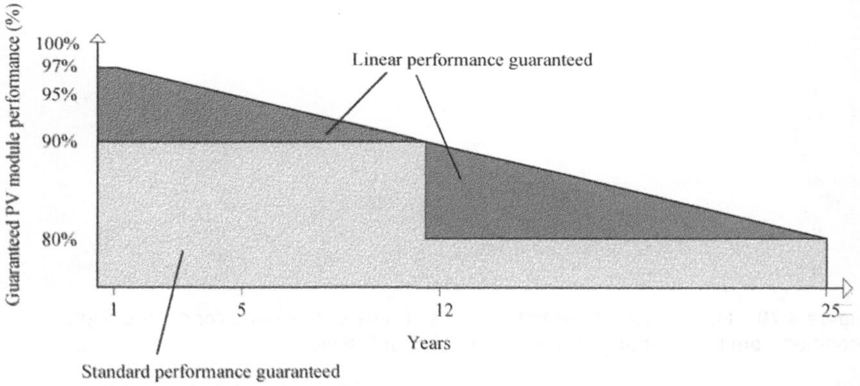
Guaranteed PV module performance (%)
100%
97%
95%
90%
80%
Linear performance guaranteed
Standard performance guaranteed
1
5
12
25
Years

Figure 8.21 PV module performance during its life time. *Source:* Based on Abdelhamid [33].

Table 8.2 Average yearly output loss of PV cells.

PV cell type	Output (%/year)
Monocrystalline silicon (mono-si)	0.36
Cadmimum telluride (CdTe)	0.4
Polycrystalline silicon (poly-Si)	0.64
Amorphous silicon (a-Si)	0.87
Copper indium gallium selendie (CIGS)	0.96

8.2.3 DC to AC Conversion Losses

8.2.3.1 Inverter Losses

An important factor that causes losses in the power generation of a PV power plant is the variable efficiency of solar inverters. The efficiency of a solar inverter is not constant and depends on the inverter load. The inverter efficiency is expressed as.

$$\eta(t) = \frac{P_{AC}(t)}{P_{DC}(t)} = \frac{P_{DC}(t) - P_{loss}(t)}{P_{DC}(t)} \tag{8.8}$$

where $P_{DC}(t)$, $P_{AC}(t)$, and $P_{loss}(t)$ are inverter DC instantaneous power, AC power, and power losses, respectively.

Inverters convert current from DC to AC at different efficiencies depending on the inverter load. Manufacturers usually provide the inverter efficiency profiles for low, medium, and high voltages. Feeding the efficiency data into modeling software provides more accurate inverter losses. The efficiency diagrams of solar inverters made by several manufacturers are presented in Figure 8.22. The inverter rated powers is in the range of 3.6–1000 kW.

Figure 8.22 shows that the efficiency of all inverters for the power levels less than 0.2 per unit decreases significantly. This means that an inverter that is much larger than the rated power operates at low efficiency. On the other hand, an inverter with a small rated power cannot transfer DC power greater than its rated value to the AC network. For a PV power plant, a PDC curve represents the total number of hours per year where the DC power is greater than a certain value. A

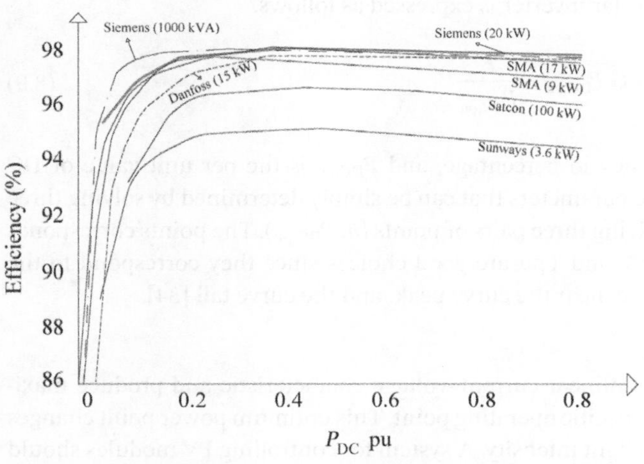
Siemens (1000 kVA)
Siemens (20 kW)
SMA (17 kW)
SMA (9 kW)
Danfoss (15 kW)
Satcon (100 kW)
Sunways (3.6 kW)
Efficiency (%)
98
96
94
92
90
88
86
0
0.2
0.4
0.6
0.8
0.8
P_{DC} pu

Figure 8.22 Efficiency curve of various solar inverters. *Source:* Modified from Malamaki and Demoulias [34].

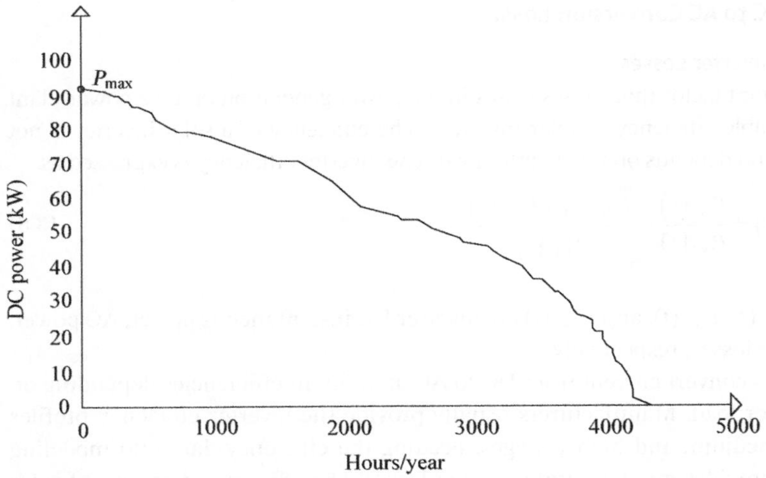
100
90
80
70
60
50
40
30
20
10
0
DC power (kW)
Pmax
0
1000
2000
3000
4000
5000
Hours/year

Figure 8.23 Typical power duration curve of 100 kWp PV installation. *Source:* Modified from Malamaki and Demoulias. [34].

typical PDC curve for a 100 kW PV plant is shown in Figure 8.23. There is an optimal inverter size that maximizes the inverter efficiency throughout its operating range. It is seen from Figure 8.23 that the maximum efficiency of an inverter is achieved only in a small range around its rated power. Therefore, the efficiency of an inverter should be evaluated in the whole operating range and not only around its maximum efficiency [34].

The efficiency of a solar inverter is expressed as follows.

$$\eta\left(P_{DC,pu}\right) = D + G.P_{DC,pu} + \frac{F}{P_{DC,pu}} \tag{8.9}$$

where η is the efficiency in percentage, and $P_{DC,pu}$ is the per unit value of DC power. D, G, and F are parameters that can be simply determined by solving three linear equations involving three pairs of points (η, $P_{DC,pu}$). The points corresponding to $P_{DC,pu} = 0.1$, 0.2, and 1 pu are good choices since they correspond to the rising front of the curve, near the curve peak, and the curve tail [34].

8.2.3.2 MPPT Losses

PV modules have a nonlinear current–voltage characteristic and produce maximum power only at a specific operating point. This optimum power point changes with temperature and light intensity. A system for controlling PV modules should operate the modules at the maximum power-producing condition, which is called MPP. If the MPP is changed due to weather conditions, the MPP controller should quickly track the maximum power point. This type of continuous tracking is called "MPPT." The MPPT is usually carried out by a DC–DC converter, which is included in most modern inverters to maximize the energy received from PV modules during their service life.

Different methods have been proposed and implemented for the MPPT. Their application in solar inverters can lead to different efficiencies and losses. The MPPT methods are divided into three general categories: offline, online, and hybrid MPPT methods.

In the offline methods, also known as model-based method, the characteristic data of PV modules are commonly used to generate the control signals by the MPPT algorithm. The offline methods of MPPT are [35]:

- Open-circuit voltage (OCV) method
- Short-circuit current (SCC) method
- Artificial intelligence (AI) method

In the online methods, also known as model-free methods, the instantaneous values of current and output voltage of PV modules are usually used by the MPPT algorithm to generate the control signals. The online methods include the following [35]:

- Perturbation and observation (P&O) method
- Extremum seeking control (ESC) method
- Incremental conductance (IncCond) method

In the hybrid methods, the control signal of the algorithm consists of two parts. Each part is made by a separate algorithmic loop. The first part can be created by one of the offline methods. This part of the control signal follows roughly the MPP and is only needed to respond quickly to the environmental changes. The second part of the control signal is created based on an online method and tracks accurately the MPP. Figure 8.24 shows a general description of the hybrid method. It shows that the first part of the control signal is generated using an offline method through the set-point calculation loop, and the second part is generated using an online method through the fine-tuning loop [35].

In a LS-PVPP, the efficiency of MPPT algorithm is one of the most important primary considerations in reducing losses. The efficiency of the MPPT algorithm is expressed as follows.

$$\eta_T = \frac{1}{n}\sum_{i=0}^{n}\frac{P_i}{P_{\max,i}} = \frac{1}{n}\sum_{i=0}^{n}1 - \frac{P_1}{P_{\max,i}} \tag{8.10}$$

where P_i is the PV module power, $P_{\max,i}$ is the module maximum power, P_1 is the power losses, and n is the number of samples. In [35], the efficiencies of different

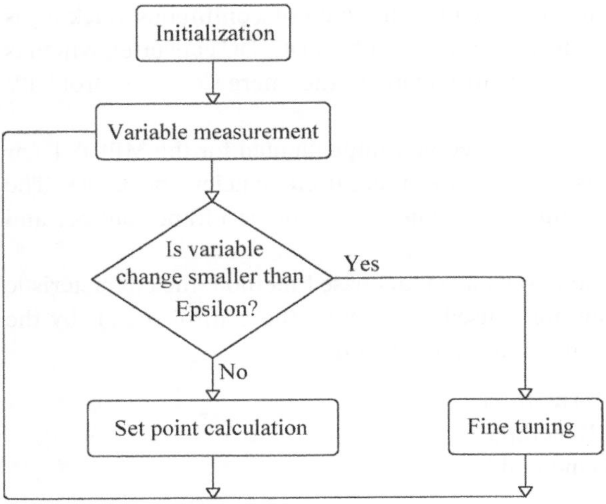
Initialization
Variable measurement
Is variable change smaller than Epsilon?
Yes
No
Set point calculation
Fine tuning

Figure 8.24 General algorithm of hybrid methods. *Source:* Reisi et al. [35]/with permission of Elsevier.

MPPT algorithms are compared. The results show that the respective efficiency obtained by each of the MPPT algorithms including the hybrid, ANN, FLM, P&O, IncCond, and ESC algorithms.

8.2.3.3 Tracking Curtailment

The construction of a LS-PVPP equipped with a solar tracking system has a significant cost for the investors. The investors expect their investment to be maintained during the useful period of 25–30 years. In this regard, one of the key components of a large-scale PV plant is the solar tracking system whose stability and resilience against storms and strong winds is important. This issue requires advanced engineering studies to identify the parameters that have the greatest contribution to the structural stability of the solar tracking system.

In the past years, major breakdowns in LS-PVPPs have occurred due to unfavorable weather conditions and strong winds. In a PV plant equipped with a tracking system, the slope angle of the PV modules changes during the day. Therefore, the wind load pressure on the PV modules varies as shown in Figure 8.25. For example, the results of wind load studies on uniaxial solar systems in Australia, the United States, and Spain show that the wind load pressure on PV modules at a slope angle of 30° is 1.8 times higher than that at a zero-slope angle. One way to protect a PV plant from fast winds is to lock the stow mode at zero degree angle. By installing a wind speed sensor, if the wind speed exceeds a threshold, the tracking system is locked at a zero degree angle, and the plant continues generating energy, while the PV modules are kept at a fixed angle. Under the fixed-angle

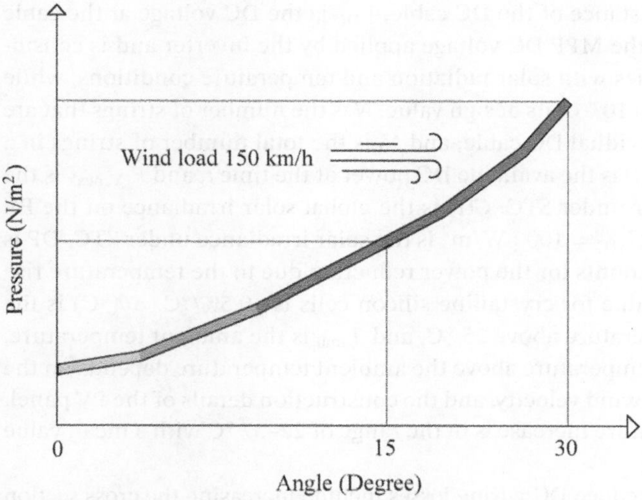
Wind load 150 km/h
Pressure (N/m²)
0
15
30
Angle (Degree)

Figure 8.25 Increasing module load with different tilt angles for Ideematec's Horizon. *Source:* Based on Abdelhamid [33].

condition, the PV modules do not operate under the optimal condition and therefore energy losses occur [36].

8.2.3.4 PV Plant DC Losses

In a PV power plant, DC cables are used to connect the PV strings to the inverter, resulting in ohmic losses due to current flow. The DC power losses of a cable at the time t is obtained from (8.11) to (8.13) [34].

$$P_{\mathrm{DC-cbl,loss}}(t) = 2 \cdot I^2_{\mathrm{DC,cable}}(t) \cdot r_{\mathrm{DC}}$$

$$= 2\left(\frac{P_{\mathrm{DC,cable}}(t)}{V_{\mathrm{DC}}}\right)^2 \cdot r_{\mathrm{DC}} = \frac{2 \cdot r_{\mathrm{DC}}}{V^2_{\mathrm{DC}}} P^2_{\mathrm{DC,cable}}(t) \tag{8.11}$$

$$= \frac{2 \cdot r_{\mathrm{DC}}}{V^2_{\mathrm{DC}}} \left(\frac{N}{N_{\mathrm{t}}}\right)^2 P^2_{\mathrm{DC}}(t)$$

$$P_{\mathrm{DC}}(t) = P_{\mathrm{PV,peak}} \frac{G(t)}{G_{\mathrm{STC}}}\left[1 + \mathrm{DP} \cdot \Delta\theta\right] \tag{8.12}$$

$$\Delta\theta = \mathrm{T_{amb}} + 30 - 25 \tag{8.13}$$

where r_{DC} is the resistance of the DC cable, V_{DC} is the DC voltage at the cable terminals, which is the MPP DC voltage applied by the inverter and is considered constant. It varies with solar radiation and temperature conditions, while its variation is within 10% of its design value. N is the number of strings that are connected to an individual DC cable, and N_t is the total number of strings in a PV installation. $P_{DC}(t)$ is the available DC power at the time t, and $P_{PV,\text{peak}}$ is the installed peak power under STC. $G(t)$ is the global solar irradiance on the PV plane at time t, and $G_{STC} = 1000\ \text{W/m}^2$ is the solar irradiance under STC. DP is a coefficient that accounts for the power reduction due to the temperature rise in cells. A typical value for crystalline silicon cells is –0.5%/°C. $\Delta\theta$(°C) is the rise of the cell temperature above 25 °C, and T_{amb} is the ambient temperature. The rise of the cell temperature above the ambient temperature depends on the solar irradiance, the wind velocity, and the construction details of the PV panel. Usually, the temperature increase is in the range of 22–37 °C with a mean value of 30°C.

The methods that reduce DC wiring losses include increasing the cross section of DC cables, changing in the wiring, and optimal arrangement of tables based on the DC cable reduction pattern. Figure 8.26 shows an example of changing the wiring strings to reduce DC cable losses. It should be noted that the size of the output cable of the modules can change to achieve an optimal DC wiring.

8.2.4 PV Plant AC Losses

8.2.4.1 AC Losses

The AC losses include the sum of LV, MV, and HV losses. To obtain AC losses, first, the losses of cables and transformers for each section are calculated. Then, the losses are added to obtain the total AC losses. For a PV power plant, the AC power dissipated in an AC cable at time t is obtained as follows [34].

$$\begin{aligned} P_{AC-cbl,loss}(t) &= 3 \cdot I_{AC,cable}^2(t).r_{AC} \\ &= 3\left(\frac{\eta(t) \cdot P_{DC,cable}(t)}{\sqrt{3} \cdot V_{LL}}\right)^2 \cdot r_{AC} \\ &= \frac{r_{AC}}{V_{LL}^2} \cdot \left(\frac{N_i}{N_{it}}\right)^2 \cdot P_{DC}^2(t) \cdot \eta^2(t) \end{aligned} \tag{8.14}$$

where r_{AC} is the AC cable resistance, $I_{AC,\text{cable}}$ is the line current rms value, and V_{LL} is the rms value of the line voltage. N_{it} is the total number of inverters, and N_i is the number of the inverters that are connected to an individual AC cable. In (8.14), it is assumed that (i) the inverter operates at the unity power factor as is the case

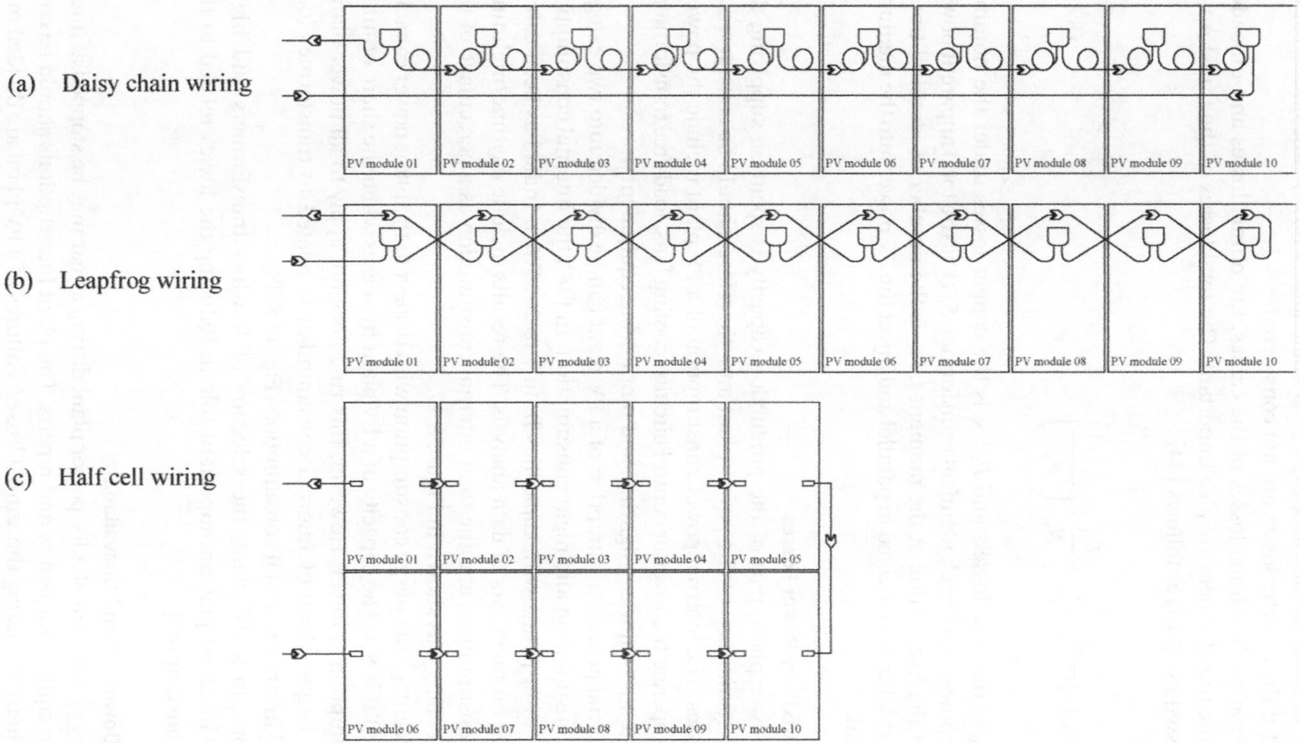
(a) Daisy chain wiring
PV module 01
PV module 02
PV module 03
PV module 04
PV module 05
PV module 06
PV module 07
PV module 08
PV module 09
PV module 10
(b) Leapfrog wiring
PV module 01
PV module 02
PV module 03
PV module 04
PV module 05
PV module 06
PV module 07
PV module 08
PV module 09
PV module 10
(c) Half cell wiring
PV module 01
PV module 02
PV module 03
PV module 04
PV module 05
PV module 06
PV module 07
PV module 08
PV module 09
PV module 10

Figure 8.26 Example of changing the wiring strings to reduce DC cable losses. (a) Daisy chain wiring, (b) Leapfrog wiring, and (c) half-cell wiring. *Source:* Based on Mahachi [37].

in almost all installations; and (ii) V_{LL} is constant. In (8.14), the value of η (t) becomes 1 if the inverter losses are not considered.

In addition to the ohmic losses of the cables, the ohmic losses and unloaded losses of the transformers are also important. The total losses of the transformer are expressed by (8.15) as follows [34].

$$P_{tr,loss}(t) = N_{tr} \cdot P_{Fe} + \frac{P_{Cu,N}}{N_{tr}} \left(\frac{S_t(t)}{S_N} \right)^2 \tag{8.15}$$

where P_{Fe} is the core losses, and $P_{Cu,\,N}$ is the copper losses under the nominal operating power S_N of each identical transformer. $S_t(t)$ is the total apparent power loading of the installation at the moment t. N_{tr} is the number of identical transformers, which are connected in parallel and inject the PV power into the medium-voltage grid.

8.2.4.2 Auxiliary Power Losses

In a PV power plant, part of the production capacity is spent on supplying the internal consumption of the power plant and, therefore, should be considered as energy losses. The internal power consumption of a PV plant include the powers needed to operate the inverter control circuits, cooling fans, load-free transformers, CCTV system, computers, lighting, and monitoring equipment.

The internal power consumption of a PV plant can be divided into two categories: day consumption and night consumption. In [38], the internal consumption of 19 PV power plants with capacities from 1 to 20 MW in India has been calculated based on two years of data analysis. The results include daytime/nighttime internal consumption, and the total internal consumption as a percentage of the total plant energy, as shown in Figure 8.27.

The percentage of energy consumption with respect to the plant power capacity is given in Table 8.3. The capacity of a PV plant increases when the share of internal consumption is lower. However, this rule does not apply to all power plants since the largest share of internal consumption is related to transformer load losses and inverter internal consumption (Figure 8.28).

Therefore, in a PV plant, the selection of low-loss transformers and high-efficiency inverters play an important role in reducing the losses related to the internal consumption.

8.2.4.3 Downtime and Unavailability

The energy production of a PV power plant during a year may be stopped at intervals due to equipment failure and repairs. The plant interruptions should be considered when calculating the annual losses. Failures of a PV plant are divided into

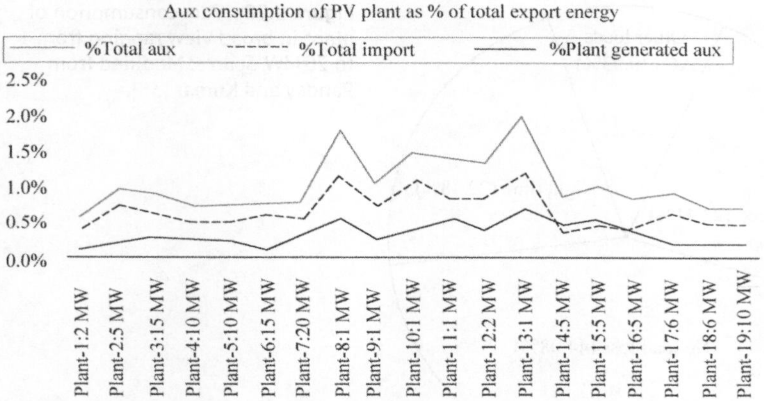
Aux consumption of PV plant as % of total export energy
%Total aux
%Total import
%Plant generated aux
2.5%
2.0%
1.5%
1.0%
0.5%
0.0%
Plant-1:2 MW
Plant-2:5 MW
Plant-3:15 MW
Plant-4:10 MW
Plant-5:10 MW
Plant-6:15 MW
Plant-7:20 MW
Plant-8:1 MW
Plant-9:1 MW
Plant-10:1 MW
Plant-11:1 MW
Plant-12:2 MW
Plant-13:1 MW
Plant-14:5 MW
Plant-15:5 MW
Plant-16:5 MW
Plant-17:6 MW
Plant-18:6 MW
Plant-19:10 MW

Figure 8.27 Aux consumption of plant in broad view ranging from 1 to 20 MW [38].

Table 8.3 Aux consumption of PV plant of various capacity [38].

Plant capacity (MW)	0–5	5–10	10–20	>20
% Aux consumption	1.54%	0.9%	0.85%	0.82%

five categories: photovoltaic, electrical, electronic, telecommunication, and civil failures as shown in Figure 8.29 [39]. Failures can be related to the modules, inverters, mounting structure, connection and distribution boxes, cabling, potential equalization and grounding, lightning and protection system, weather station, communication and monitoring, transformer station, infrastructure and environmental influence, storage system, etc.

During the operation/maintenance phase of PV plant, failures can be found in the PV array. The failures include snail trail, hot spot, diode failure, EVA discoloration, glass breakage, delamination with breaks in the ribbons and solder bonds, LID, low irradiance losses, PID, shading effect, soiling effect, sun tracking system misalignments, wiring losses, mismatching effect in solar array, ground faults, line-to-line faults, and arc faults. Although there have not been many such failures in the existing PV power plants, a recent study indicates the need for improvements to avoid potential failures [39].

In [40], statistical data related to the failure of equipment of 63 PV power plants located in Italy and Spain with a capacity of 200 kW to 10 MW for a period of five years (2012–2016) have been collected and processed. The distribution diagram of the failure percentage of the equipment is presented in Figure 8.30. It can be

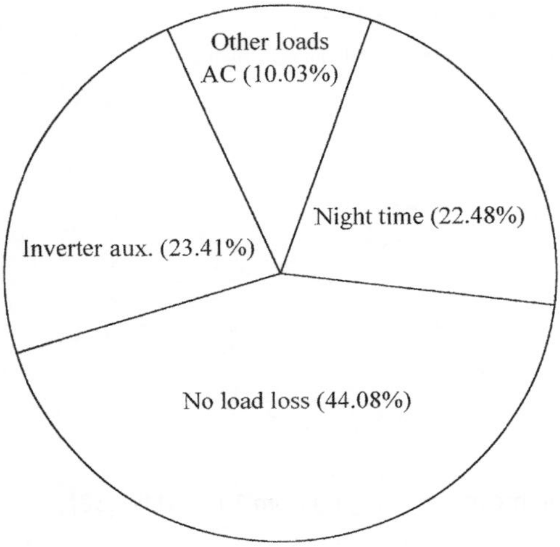
Other loads
AC (10.03%)
Night time (22.48%)
Inverter aux. (23.41%)
No load loss (44.08%)

Figure 8.28 Aux consumption of plant in broad view ranging from 1 to 20 MW. *Source:* Modified from Pandey and Kumar [38].

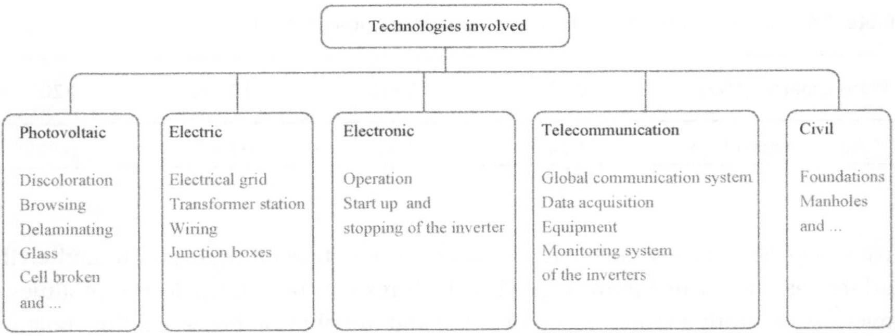
Technologies involved
Photovoltaic
Discoloration
Browsing
Delaminating
Glass
Cell broken
and ...
Electric
Electrical grid
Transformer station
Wiring
Junction boxes
Electronic
Operation
Start up and
stopping of the inverter
Telecommunication
Global communication system
Data acquisition
Equipment
Monitoring system
of the inverters
Civil
Foundations
Manholes
and ...

Figure 8.29 Classification of failures according to the involved technology. *Source:* Modified from Lillo-Bravo et al. [39].

observed that the monitoring system, telecommunications, inverter, power network, and PV generator had the largest share in the recorded failures.

Various reliability indicators are provided for the analysis of failure statistics. One of these indicators is the calculation of the total number of failures of the PV power plant per kilowatt of installed capacity, which is presented in Table 8.4. The average of this index showing the total failure number without considering the capacity of power plants is equal to 0.23.

The studied PV plants are classified based on their capacity and the number of failures per kilowatt as shown in Figure 8.31. It is observed that for the plant capacity less than 750 kW, the failure index is 0.45 and for the plant capacity above 750 kW, the failure index is equal to 0.18. Another important reliability index is the average repair time of equipment per failure, which is shown in Figure 8.32.

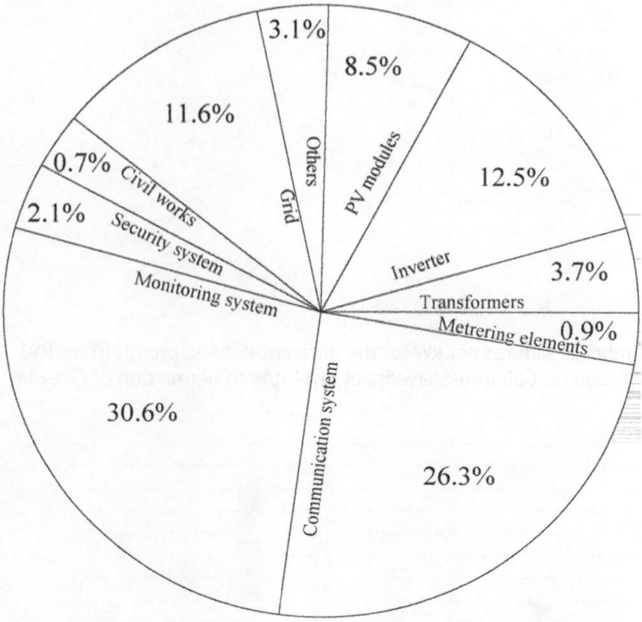
3.1%
8.5%
11.6%
0.7%
Civil works
2.1%
Security system
Monitoring system
Others
Grid
PV modules
12.5%
Inverter
3.7%
Transformers
Metrering elements
0.9%
Communication system
30.6%
26.3%

Figure 8.30 Distribution of the registered failures in each of the PV plants monitored elements, considering data from 2012 to 2016. *Source:* Gallardo-Saavedra et al. [40]/with permission of Elsevier.

Table 8.4 PV plants annual failure ratio per kW [40].

Year	Failure(kW)
2012	0.23
2013	0.24
2014	0.21
2015	0.28
2016	0.21
Average	0.23

Source: Gallardo-Saavedra et al. [40]/with permission of Elsevier.

The results show that the safety system, the civil works, and the monitoring system have the longest repair time.

8.2.4.4 Grid Compliance Losses

Overloading of local transmission or distribution equipment such as overhead lines or power transformers may lead to network instability. In such cases, the

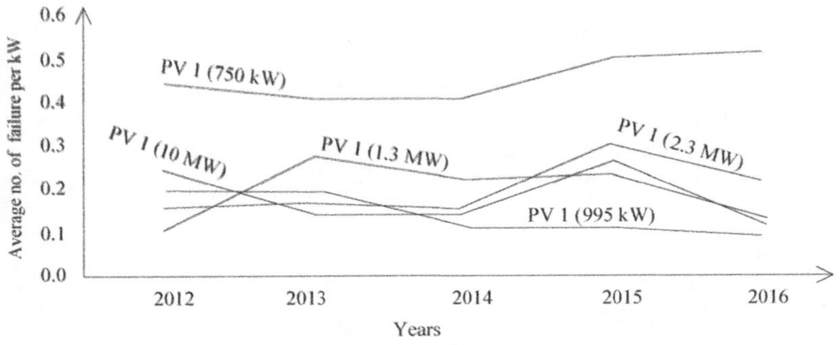
0.6
0.5
0.4
0.3
0.2
0.1
0.0
Average no. of failure per kW
PV 1 (750 kW)
PV 1 (10 MW)
PV 1 (1.3 MW)
PV 1 (2.3 MW)
PV 1 (995 kW)
2012
2013
2014
2015
2016
Years

Figure 8.31 Average number of failures per kW for the different defined groups (from PV1 to PV5) from 2012 to 2016. *Source:* Gallardo-Saavedra et al. [40]/with permission of Elsevier.

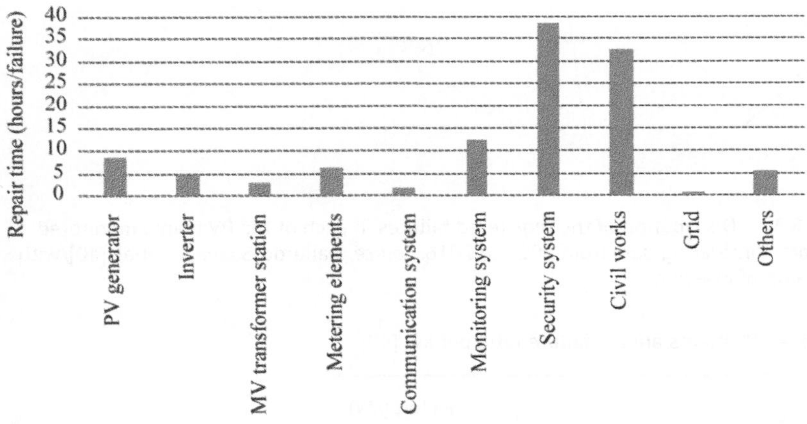
Repair time (hours/failure)
40
35
30
25
20
15
10
5
0
PV generator
Inverter
MV transformer station
Metering elements
Communication system
Monitoring system
Security system
Civil works
Grid
Others

Figure 8.32 Reparation time per failure for different PV plant components. *Source:* Gallardo-Saavedra et al. [40]/with permission of Elsevier.

grid voltage and frequency may be out of range of the inverters and the PV plant interruption may occur. In less-developed regional networks, the risk of failure due to network instability can have serious effects on the annual plant revenue.

8.3 Energy Yield Prediction

An average annual energy estimate over the useful life of a PV power plant, which is between 25 and 30 years, is required to calculate the plant revenue. For this purpose, energy yield analysis is performed to predict the energy expected from

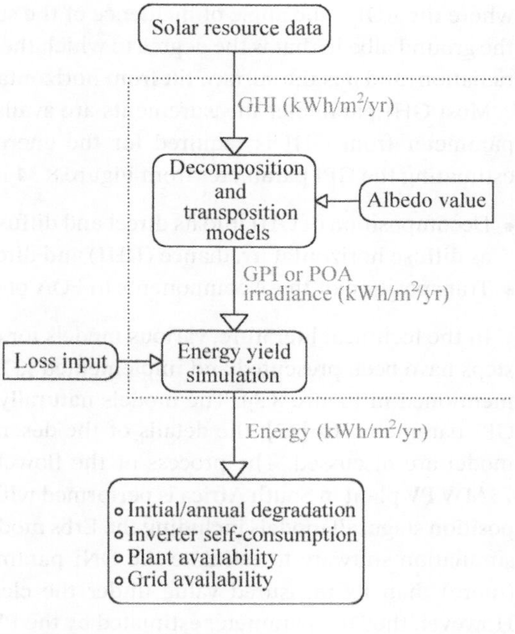

Solar resource data
GHI (kWh/m²/yr)
Decomposition and transposition models
Albedo value
GPI or POA irradiance (kWh/m²/yr)
Loss input
Energy yield simulation
Energy (kWh/m²/yr)
o Initial/annual degradation
o Inverter self-consumption
o Plant availability
o Grid availability

Figure 8.33 The procedure for predicting the energy yield of a PV plant [37].

the PV plant. This analysis, according to the flowchart shown in Figure 8.33, includes the following steps [37].

- Estimation of radiation on the surface of PV modules
- Estimation of PV plant losses
- Modeling and simulation considering the model uncertainty and inputs
- Estimation of annual energy yield and calculation of performance ratio (PR) and capacity factor (CF) indices

8.3.1 Irradiation on Modules

The first step in energy yield analysis is to estimate the total radiation on the surface of PV modules, i.e. POA or GPI. This includes three components as direct radiation, diffused radiation, and reflected radiation from the ground. The relationship between the POA and the three components of radiation is expressed by the following equations [41],

$$\mathrm{POA}(\mathrm{GPI}) = G_{\mathrm{Dir}} + G_{\mathrm{diff}} + G_{\mathrm{Ref}} \tag{8.16}$$

$$G_{\mathrm{Dir}} = \mathrm{DNI} \times \cos(\mathrm{AOI}) \tag{8.17}$$

$$G_{\mathrm{Ref}} = \mathrm{GHI} \times \rho \times \frac{1-\cos(\beta)}{2} \tag{8.18}$$

where the AOI is the angle of incidence of the sun beam on the POA surface, ρ is the ground albedo that is the degree to which the ground is able to reflect diffusely radiation, and β is the surface tilt from horizontal.

Most GHI parameter measurements are available. However, extraction of GPI parameter from GHI is required for the energy yield analysis. Modeling and estimating the GPI parameter from Figure 8.34 includes:

- Decomposition of GHI into its direct and diffuse components, usually expressed as diffuse horizontal irradiance (DHI) and direct normal irradiance (DNI)
- Transposition of these components to POA of the modules

In the technical literature, various models for decomposition and transposition steps have been presented and implemented [37, 39]. Some of these methods are mentioned in Figure 8.34. The models naturally lead to various amounts for the GPI parameter. In [37], the details of the description and relationships of each model are discussed. The process of the flowchart shown in Figure 8.34 for a 75 MW PV plant in South Africa is performed with various models. For the decomposition stage, all models including the Erbs model are implemented in the PVsyst simulation software to estimate the DNI parameter. The DNI parameter is less (more) than its measured value under the clear (cloudy) weather conditions. However, the DHI parameter estimated by the PVsyst software is more (less) than its measured value under the clear (cloudy) weather conditions. The results for the Erbs model for the two conditions of the clear air and cloudy weather are shown in Figure 8.35. For the transposition stage, Perez model presented in Chapter 5 estimates the GPI parameter more accurately than the other models.

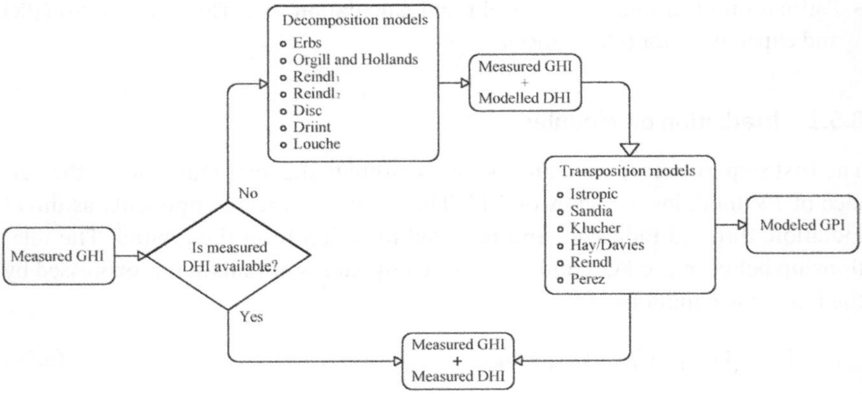
Decomposition models
Erbs
Orgill and Hollands
Reindl1
Reindl2
Disc
Driint
Louche
Measured GHI
+
Modelled DHI
Transposition models
Istropic
Sandia
Klucher
Hay/Davies
Reindl
Perez
Modeled GPI
Measured GHI
Is measured DHI available?
No
Yes
Measured GHI
+
Measured DHI

Figure 8.34 Flowchart showing how to model GPI from measured GHI [37].

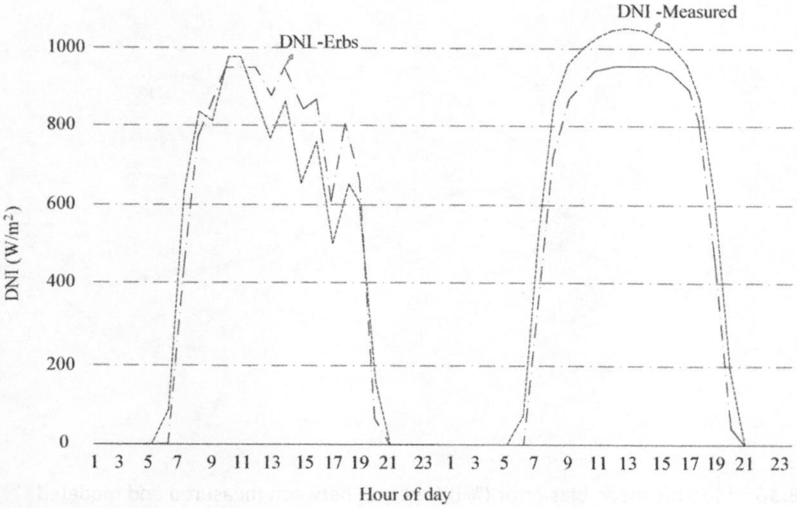
DNI -Measured
DNI -Erbs
1000
800
600
400
200
0
DNI (W/m²)
1 3 5 7 9 11 13 15 17 19 21 23 1 3 5 7 9 11 13 15 17 19 21 23
Hour of day

Figure 8.35 Estimation of DNI using Erbs decomposition model on a cloudy day and on a clear day [37].

The final results show that the combination of DIRINT model from decomposition stage with Perez model from transposition stage has a prediction error of less than 2% for the GPI parameter as shown in Figure 8.36.

8.3.2 PV Plant Losses

One of the important factors for estimating the average annual energy of a PV power plant is modeling the plant losses. Various equipment and their conditions lead to energy losses, which are described in detail in Section 8.2.

8.3.3 Performance Modeling

To evaluate the efficiency of a PV power plant and to calculate the annual energy under various environmental conditions and operations, modeling of the plant by simulation tools is required. The PVsyst tool is one of the most well-known simulation software packages in this field. This software receives the site information as input, e.g. GHI information, its hourly, daily and monthly amounts, temperature, losses, albedo coefficient, and calculates the expected annual energy. The performance of a software package in estimating the annual energy is evaluated using the measurement data.

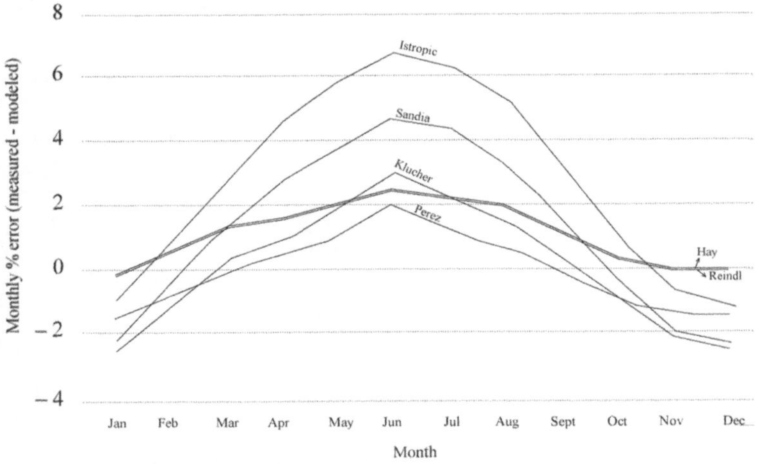
8
6
4
2
0
−2
−4
Monthly % error (measured - modeled)
Istropic
Sandia
Klucher
Perez
Hay
Reindl
Jan
Feb
Mar
Apr
May
Jun
Jul
Aug
Sept
Oct
Nov
Dec
Month

Figure 8.36 Monthly mean bias error (% difference) between measured and modeled GPI using the Dirint decomposition model [37].

8.3.4 Uncertainty in Energy Yield

The process of modeling and preparing input parameters for simulation software is always faced with uncertainties. Modeling by a software itself has an uncertainty of 2–3%. Measurement accuracy in meteorological stations is 3–5%. Uncertainties in the calculation of energy losses such as those related to the power outages and dust pollution are also considerable. In general, modeling and parameter uncertainties can reach up to 10% and their calculations require a more detailed analysis of the site [42].

8.3.5 Performance Ratio

One of the most important parameters for evaluating the efficiency of a PV power plant is the performance ratio (PR) index, which is usually expressed in percent. The PR index is defined in IEC 61724 [43] and depends on the climatic conditions and system losses. To increase the power plant PR, the system losses should be reduced as much as possible using an optimal design. The higher the PR of a PV plant, the more efficient the plant is in terms of converting solar radiation into useful energy.

Since the losses of a PV plant vary according to the environmental conditions and at different times, the plant PR varies over time too. This index calculates the ratio between the actual produced energy and the nominal energy yield. The pure energy without any losses is calculated as [44].

$$PR = \frac{\text{Actual yield in}\left(\text{kWh}\right)}{\text{Calculated nominal yield}} \tag{8.19}$$

$$\text{Calculated nominal yield} = \text{GHI}\left(\text{in kWh / m}^2\right) \times \text{Rated module efficiency} \times \text{Total PV area}\left(\text{in m}^2\right) \tag{8.20}$$

As an example, the values of parameters for an under-study PV plant are given below [45].

- Analysis period: one year
- Measured average solar irradiation intensity in one year: $120\,\text{kWh/m}^2$
- Generation area of the PV plant: $10\,\text{m}^2$
- Efficiency factor of the PV modules: 15%

  Electrical energy exported by the plant to the grid: 110 kWh
  Calculated nominal yield = $120\,\text{kWh/m}^2 \times 15\% \times 10\,\text{m}^2 = 180\,\text{kWh}$
  The PR value is then calculated as follows.
  PR = $110/180 \times 100 = 61.1\%$

The PR value of 61% indicates that about 39% of the solar energy in the analysis period is not converted into the usable energy due to the circumstances such as conduction losses, thermal losses, or defects in the components. The performance ratio can be used as an indicator whether or not a more detailed inspection of the PV plant is required. For PV plant with low PR, the level of soiling of the PV modules should be verified and if needed, it should be removed. Moreover, the defective components should be repaired or replaced.

According to the National Renewable Energy Laboratory (NREL), for a newly built power plants, the PR index is expected to be at least 77% with an annual decrease of about 1% [37]. In [46], the PR index for a 6.4 kW solar system in three geographical areas in Greece, Germany, and England has been calculated for the fixed angle and biaxial systems. Based on the obtained results, the rates of improvement of PR index for the two-axis system in Greece, Germany, and England are 0.8, 1.1, and 1.3%, respectively.

8.3.6 Capacity Factor

The capacity factor (CF) index is the ratio of the measured energy produced in a certain time period, and the energy capacity at the rated power of the plant for the same period. The value of the CF index for a period of one year is expressed as [46].

$$CF = \frac{\text{Energy Output}\,(\text{MWh})}{365 \times 24 \times \text{installed capacity of the PV system}\,(\text{MWp})} \tag{8.21}$$

The CF of a fixed tilt PV plant in the Southern Spain is around 16%. This means that a 5 MW PV power plant generates the equivalent energy of a continuously operating 0.8 MW conventional power plant [42]. In the calculation of the CF index, the environmental changes, e.g. temperature, radiation, and panel failure, are not included. In addition, the plant is assumed to generate 24 hours a day. According to some studies, e.g. [37, 42], the expected value of the CF index in the areas with high radiation is between 20.8 and 26%.

8.4 Conclusion

The PV power plants are advancing in terms of their technology and efficiency. However, the breakdowns and energy losses associated with the equipment are also growing rapidly, and new issues are emerging. As explained in this chapter, many factors affect the power losses and output energy of a PV power plant. The power losses and derating of a PV can be due to temperature rise, dust, wiring failures, shading, ground slope, azimuth tracking failure, and grid outages.

For a PV power plant, the total performance ratio between 70 and 90% is acceptable. The higher the total losses of a PV power plant, the amounts of yearly energy production and revenue are lower and the payback period becomes longer.

The team working on the design of a PV power plant has the task to reduce the total plant losses as much as possible. In addition to the plant site location and weather conditions, the design of internal system of a PV plant has an important role in the overall plant losses and outages.

References

1 Martin, N. and Ruiz, J.M. (2002). A new model for PV modules angular losses under field conditions. *International Journal of Solar Energy 22* (1): 19–31.

2 Shah, A. (ed.) (2020). *Solar Cells and Modules*. Springer International Publishing.

3 Verma, A. and Singhal, S. (2015). Solar PV performance parameter and recommendation for optimization of performance in large scale grid connected solar PV plant—case study. *Journal of Energy Power Sources 2* (1): 40–53.

4 Martin, N. and Ruiz, J.M. (2001). Calculation of the PV modules angular losses under field conditions by means of an analytical model. *Solar Energy Materials and Solar Cells 70* (1): 25–38.

5 García, M., Marroyo, L., Lorenzo, E., and Pérez, M. (2011). Soiling and other optical losses in solar-tracking PV plants in navarra. *Progress in Photovoltaics: Research and Applications 19* (2): 211–217.

6 Sweerts, B., Pfenninger, S., Yang, S. et al. (2019). Estimation of losses in solar energy production from air pollution in China since 1960 using surface radiation data. *Nature Energy 4* (8): 657–663.

7 Li, X., Wagner, F., Peng, W. et al. (2017). Reduction of solar photovoltaic resources due to air pollution in China. *Proceedings of the National Academy of Sciences 114* (45): 11867–11872.

8 Abd-Elhady, M.S., Zayed, S.I.M., and Rindt, C.C.M. (2011). Removal of dust particles from the surface of solar cells and solar collectors using surfactants. In: *International Conference on Heat Exchanger Fouling and Cleaning*, vol. 5, no. 10. www.heatexchanger-fouling.com.

9 Kaldellis, J.K. and Kapsali, M. (2011). Simulating the dust effect on the energy performance of photovoltaic generators based on experimental measurements. *Energy 36* (8): 5154–5161.

10 Simpson, L.J., Muller, M., Deceglie, M. et al. (2017). NREL efforts to address soiling on PV modules. In: *2017 IEEE 44th Photovoltaic Specialist Conference (PVSC)*, 2789–2793. IEEE.

11 Konyu, M., Ketjoy, N., and Sirisamphanwong, C. (2020). Effect of dust on the solar spectrum and electricity generation of a photovoltaic module. *IET Renewable Power Generation 14* (14): 2759–2764.

12 Freeman, J.M. and Ryberg, D.S. (2017). Integration, Validation, and Application of a PV Snow Coverage Model in SAM (No. NREL/TP-6A20-68705). Golden, CO, USA: National Renewable Energy Lab (NREL).

13 Louwen, A., de Waal, A.C., Schropp, R.E. et al. (2017). Comprehensive characterisation and analysis of PV module performance under real operating conditions. *Progress in Photovoltaics: Research and Applications 25* (3): 218–232.

14 Santiago, I., Trillo-Montero, D., Moreno-Garcia, I.M. et al. (2018). Modeling of photovoltaic cell temperature losses: a review and a practice case in South Spain. *Renewable and Sustainable Energy Reviews 90*: 70–89.

15 Makrides, G., Zinsser, B., Phinikarides, A. et al. (2012). Temperature and thermal annealing effects on different photovoltaic technologies. *Renewable Energy 43*: 407–417.

16 Głuchy, D., Kurz, D., and Trzmiel, G. (2014). Analysis of the influence of shading by horizon of PV cells on the operational parameters of a photovoltaic system. *Przegląd Elektrotechniczny 90* (4): 78–80.

17 Hanson, A.J., Deline, C.A., MacAlpine, S.M. et al. (2014). Partial-shading assessment of photovoltaic installations via module-level monitoring. *IEEE Journal of Photovoltaics 4* (6): 1618–1624.

18 Satpathy, P.R. and Sharma, R. (2020). Reliability and losses investigation of photovoltaic power generators during partial shading. *Energy Conversion and Management 223*: 113480.

19 Kerekes, T., Koutroulis, E., Séra, D. et al. (2012). An optimization method for designing large PV plants. *IEEE Journal of Photovoltaics 3* (2): 814–822.

20 Sun, Y., Chen, S., Xie, L. et al. (2014). Investigating the impact of shading effect on the characteristics of a large-scale grid-connected PV power plant in Northwest China. *International Journal of Photoenergy* 2014.

21 Dolara, A., Lazaroiu, G.C., Leva, S., and Manzolini, G. (2013). Experimental investigation of partial shading scenarios on PV (photovoltaic) modules. *Energy* 55: 466–475, ISSN 0360-5442, https://doi.org/10.1016/j.energy.2013.04.009.

22 Sauer, K.J., Roessler, T., and Hansen, C.W. (2014). Modeling the irradiance and temperature dependence of photovoltaic modules in PVsyst. *IEEE Journal of Photovoltaics 5* (1): 152–158.

23 Kivalov, S.N. and Fitzjarrald, D.R. (2018). Quantifying and modelling the effect of cloud shadows on the surface irradiance at tropical and midlatitude forests. *Boundary-layer Meteorology 166* (2): 165–198.

24 Lappalainen, K. and Valkealahti, S. (2017). Effects of PV array layout, electrical configuration and geographic orientation on mismatch losses caused by moving clouds. *Solar Energy 144*: 548–555.

25 Wolny, F., Weber, T., Müller, M., and Fischer, G. (2013). Light induced degradation and regeneration of high efficiency Cz PERC cells with varying base resistivity. *Energy Procedia 38*: 523–530.

26 Vidyanandan, K.V. (2017). An overview of factors affecting the performance of solar PV systems. *Energy Scan* 27 (28): 216.

27 Woyte, A. and Goy, S. (2017). Large grid-connected photovoltaic power plants: best practices for the design and operation of large photovoltaic power plants. In: *The Performance of Photovoltaic (PV) Systems*, 321–337. Woodhead Publishing.

28 Gasparin, F.P., Bühler, A.J., Rampinelli, G.A., and Krenzinger, A. (2016). Statistical analysis of I–V curve parameters from photovoltaic modules. *Solar Energy 131*: 30–38.

29 Pavan, A.M., Tessarolo, A., Barbini, N. et al. (2015). The effect of manufacturing mismatch on energy production for large-scale photovoltaic plants. *Solar Energy 117*: 282–289.

30 Han, C. and Lee, H. (2018). Investigation and modeling of long-term mismatch loss of photovoltaic array. *Renewable Energy 121*: 521–527.

31 Sharma, V. and Chandel, S.S. (2013). Performance and degradation analysis for long term reliability of solar photovoltaic systems: a review. *Renewable and Sustainable Energy Reviews 27*: 753–767.

32 Vázquez, M. and Rey-Stolle, I. (2008). Photovoltaic module reliability model based on field degradation studies. *Progress in Photovoltaics: Research and Applications 16* (5): 419–433.

33 Abdelhamid, M.M. (2014). A comprehensive assessment methodology based on life cycle analysis for on-board photovoltaic solar modules in vehicles. Doctoral dissertation, Clemson University.

34 Malamaki, K.N.D. and Demoulias, C.S. (2013). Minimization of electrical losses in two-axis tracking pv systems. *IEEE Transactions on Power Delivery 28* (4): 2445–2455.

35 Reisi, A.R., Moradi, M.H., and Jamasb, S. (2013). Classification and comparison of maximum power point tracking techniques for photovoltaic system: a review. *Renewable and Sustainable Energy Reviews 19*: 433–443.

36 Nathalie K., 2020. Single-axis solar tracker stowing strategies: what will best protect your investment from extreme wind events, p. 11.

37 Mahachi, T. (2016). Energy yield analysis and evaluation of solar irradiance models for a utility scale solar PV plant in South Africa. Doctoral dissertation, Stellenbosch University, Stellenbosch.

38 Pandey, S. and Kumar, R. (2017). Analysis of auxiliary energy consumption in utility scale solar pv power plant. *International Journal of Current Engineering and Technology* 7 (5): 1728–1729.

39 Lillo-Bravo, I., González-Martínez, P., Larrañeta, M., and Guasumba-Codena, J. (2018). Impact of energy losses due to failures on photovoltaic plant energy balance. *Energies* 11 (2): 363.

40 Gallardo-Saavedra, S., Hernández-Callejo, L., and Duque-Pérez, O. (2019). Quantitative failure rates and modes analysis in photovoltaic plants. *Energy 183*: 825–836.

41 Lave, M., Hayes, W., Pohl, A., and Hansen, C.W. (2015). Evaluation of global horizontal irradiance to plane-of-array irradiance models at locations across the United States. *IEEE Journal of Photovoltaics 5* (2): 597–606.

42 Lumby, B. and Miller, A.J.H. (2011). Utility Scale Solar Power Plants: A Guide for Developers and Investors (No. 66762), 1–204. The World Bank.

43 IEC-International Electrotechnical Commission (2016). Photovoltaic system performance – part 2: capacity evaluation method. IEC TS 61724-2.

44 Vasisht, M.S., Srinivasan, J., and Ramasesha, S.K. (2016). Performance of solar photovoltaic installations: effect of seasonal variations. *Solar Energy 131*: 39–46.

45 Khalid, A.M., Mitra, I., Warmuth, W., and Schacht, V. (2016). Performance ratio–crucial parameter for grid connected PV plants. *Renewable and Sustainable Energy Reviews 65*: 1139–1158.

46 Axaopoulos, P.J. and Fylladitakis, E.D. (2013). Energy and economic comparative study of a tracking vs. a fixed photovoltaic system in the northern hemisphere. *International Journal of Energy, Environment and Economics* 21 (1): 1.

Index

Step-by-Step Design of Large-Scale Photovoltaic Power Plants, First Edition. Davood Naghaviha, Hassan Nikkhajoei, and Houshang Karimi.

Printed and bound by CPI Group (UK) Ltd, Croydon, CR0 4YY
16/04/2025
14658344-0001